THE AGE OF THE SOLAR SYSTEM: A STUDY OF THE POYNTING-ROBERTSON EFFECT AND EXTINCTION OF INTERPLANETARY DUST

by

Harold S. Slusher, M.S., D.Sc.

and

Stephen J. Duursma, M.S.

ICR Technical Monograph No. 6

Institute for Creation Research

San Diego, Ca 92116

ISBN - 0-932766-01-3

Library of Congress Card Catalog Number 78-70529

Institute for Creation Research

San Diego, California

Printed in the United States of America

TABLE OF CONTENTS

PART I

INTRODUCTION

The purpose of this monograph is to review some of the significant studies of the effects of solar radiation on a dust particle in interplanetary space in order to establish an upper limit on the age of the solar system. Since the authors believe the Poynting-Robertson effect plays a major role in any attempts at determining the age of the solar system, a secondary goal of this monograph is to provide a complete and detailed treatment of this problem, rather than the bare outline and results found in published works on this topic.

The Poynting-Robertson effect will be considered in detail and combined with the influence of solar radiation pressure and gravitational attraction. The results of this theoretical investigation are then applied to common sizes, densities, and orbits of the interplanetary medium as determined from experimental and observational techniques. Estimations of times required for elimination of various particles from the solar system have been made. General graphs are provided to enable a direct reading of time of infall for any particle of interest given the intial orbital parameters. These graphs are limited to particles of the range in which the Poynting-Robertson effect will be a significant dynamic influence.

PART II

INTERPLANETARY DUST

In order to become familiarized with the physical characteristics of the interplanetary dust cloud, the following brief review of experimental and observational techniques has been included. The generally accepted conclusions from these studies are also presented and will be used later on. Since the methods used to arrive at these values are not within the scope of this work, the basic results will simply be presented without a demonstration of their validity. Sources will be cited for confirmation of this information.

DATA COLLECTION METHODS

ZODIACAL LIGHT

The zodiacal light, which was first mentioned by Cassini in 1685, can be observed as a faint glow near the horizon along the ecliptic a few hours before sunrise or after sunset. One of its primary characteristics is its light spectrum which is essentially the same as that of the sun. This leads to the conclusion that the light observed is reflected sunlight from a cloud of dust particles or gas molecules. The symmetry of the dust cloud about the azimuth has been well established as well as the degree of polarization and its distribution. Other characteristics have been calculated. The absolute brightness and shape of the cloud have been used to determine the density of the dust cloud. The accuracy of this method has been the object of much debate. Determinations of size distributions, albedos, mass density and distribution in space have all been attempted and involve complicated derivations whose solutions fit a fairly wide range of these physical parameters.[1]

CORONA

It has been postulated that the corona is an inward extension of the zodiacal light since its spectrum is also like that of the sun. Measurements of relative brightness vs. elongation when plotted on a log-log graph show a correlation which seems to support this idea.[2]

GEGENSCHEIN

Another glow in the sky is found about 3° west of the antisolar point, but it is far less distinct than the zodiacal light. The gegenschein, as it is called, has a color redder than sunlight and is detectable as far as 20° from its center. There are several theories on the cause of this glow. The physics of the three-body problem has been applied to show that there could be a group of small bodies reflecting sunlight in this region. It has also been speculated that there is a tail of gases following the earth and sheltered from solar radiation. One other possibility is that this is zodiacal light which is observable because of a good illumination angle of the particles in that area of the sky.[3] Though an interesting phenomenon, the gegenschein adds little to current

---

[1]John C. Brandt and Paul W. Hodge, *Solar System Astrophysics*, (New York, 1967), pp. 294-299.

[2]*Ibid.*, pp. 298.

[3]*Ibid.*, pp. 310-311.

knowledge of the interplanetary dust cloud.

METEORS AND METEORITES

Meteors and meteorites have provided another form of evidence of the dust cloud. Photographic studies of meteors involving two or more cameras in different locations have enabled astronomers to make very accurate calculations of height, path, duration, brightness and velocity of bright meteors. Meteor spectra have also been studied to determine chemical composition. By chance, the use of radio waves to study the atmosphere in 1931 led to the discovery that sudden increases of electron densities in the E region corresponded to the incidence of meteor showers in the atmosphere. Distances to meteor paths are now routinely calculated from radio data. Accumulation of data simultaneously from three stations allows calculations of path, velocity and orbit if the path is quite long. Doppler shifts are also used to calculate meteor velocities. Both optical and radio data give information concerning the origin of the meteors if their orbits are plotted backward into space.[4] Meteor showers have been linked with the paths of comets by this means, and this is one reason comets are considered by some to be the major source of dust in the solar system.[5]

Direct study of interplanetary dust can be performed on meteoritic dust collected by various means. Ice cores in the Arctic and Antarctic

---

[4]*Ibid.*, pp. 242-248.

[5]Fred L. Whipple, "On Maintaining the Meteoritic Complex," in *The Zodiacal Light and the Interplanetary Medium*, ed. J.L. Weinberg, (Washington, D.C., 1967), pp. 420-423.

make it possible to study the accumulation of meteoritic dust which fell as much as 750 years ago. Deep-sea cores are also a source of meteoritic dust. Collections of dust are made by jets at high altitudes and by traps laid out at high altitudes on the surface of the earth. These studies are hampered by some alteration which takes place as the meteorites fall through the atmosphere, but valuable information concerning the chemical and mineral make-up of meteorites has been obtained.[6]

ROCKETS AND SATELLITES

Rocket and satellite experiments have been performed to determine mass, size, spatial distribution, and space density. Besides studies which have been made of meteoroid impacts on space capsules, there are five basic types of rocket collectors which provide meteorite data. They are piezo-electric microphone systems, penetration detectors, capacitor detectors, ionization detectors, and semiconductor detectors. Of these the ionization detector is the most important and most often used. A high velocity dust particle produces a plasma on impact with a solid target and the amplitude of the resulting signal is measured and used to determine the particle mass.[7]

LUNAR METHODS

One further method has been used to obtain data concerning the

---

[6]Brandt and Hodge, pp. 305-307.

[7]H. Fechtig, "In-Situ Records of Interplanetary Dust Particles--Methods and Results," in *Interplanetary Dust and Zodiacal Light*, eds. Hans Elsässer and H. Fechtig (New York, 1976), pp. 143-144.

interplanetary medium. The seismographs set up on the moon and study of moon craters are used to study impacts of objects with mass > 10^2 *g*. The study of microcraters on lunar rock samples is used to determine mass and density of particles from 10^{-8} to 10^{-14} *g*.[8]

PHYSICAL PARAMETERS

Table I is a compilation of generally accepted averages of physical parameters of interplanetary dust. These are calculated using data from many of the methods listed above.

TABLE I

Properties of Interplanetary Dust

Physical:	
Size range (μ)	1-300
Size spectrum	s^{2-6} ds
Physical density (g/cm^3)	3.5
Composition	Stoney
Mass concentration (g/cm^3)	3×10^{-21}
Shape	Near Spherical
Surface albedo	0.1
Phase functions (mag/deg)	0.02
Orbital:	
Spatial variations	$r^{-1.5}$
Geocentric velocity (km/sec)	5^{-15}
Semi-major axis	$a^{0.5}$
Eccentricity	<0.5
Inclination	≃30°

(After Singer)

Further general information has been reviewed by Peter Millman in

[8]Peter M. Millman, "Dust in the Solar System," in *The Dusty Universe*, eds. George B. Field and A.G.A. Cameron (New York, 1975), p. 187.

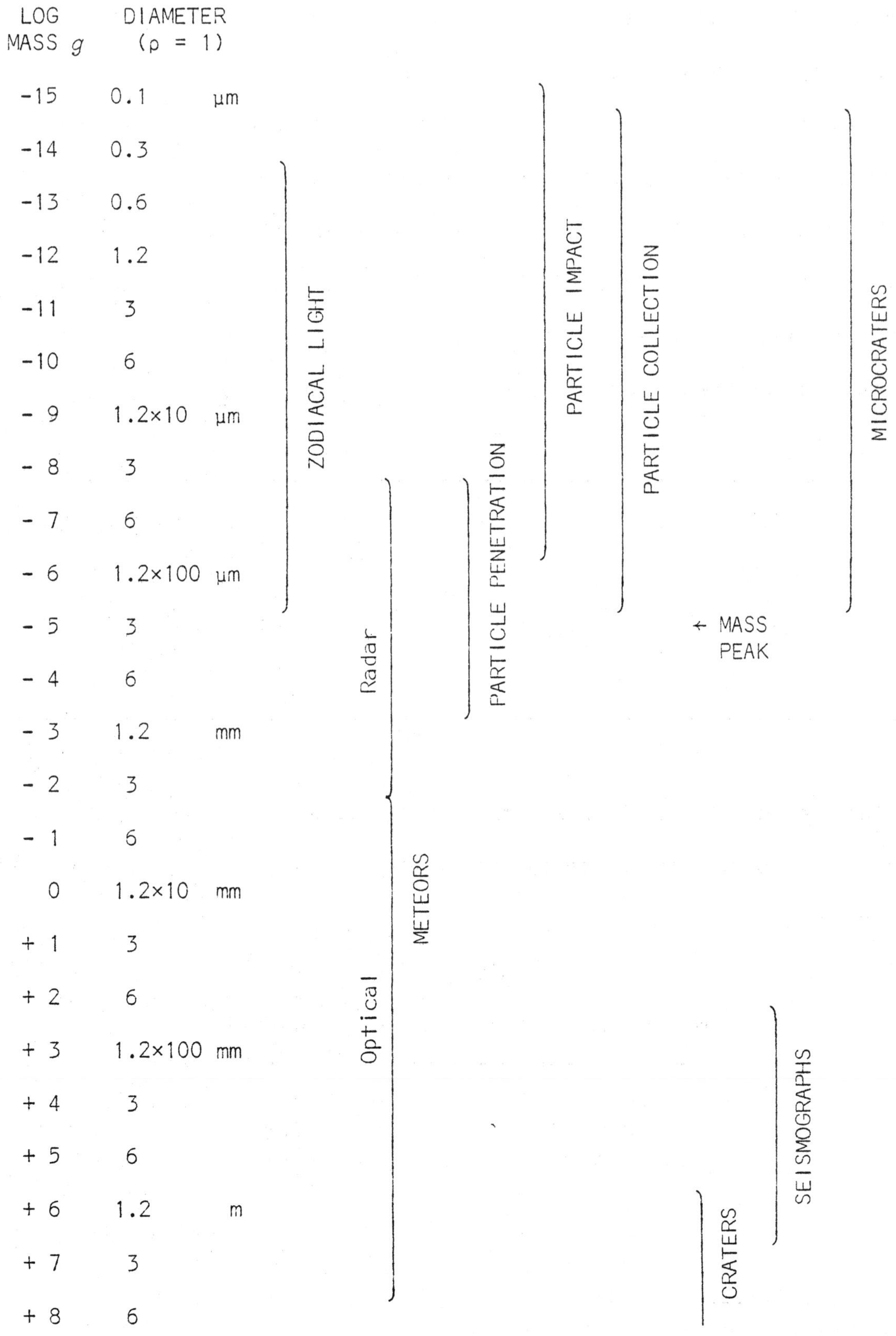

Figure (1). Techniques and applicable mass ranges.

figure (1) which shows the ranges of applicability for collecting data by many of the methods discussed above.

Other pertinent data with respect to this thesis deal with the percentages of the total mass of interplanetary dust which fall into certain ranges and with flux variations with solar distance. Millman finds that two thirds of the total mass of the dust cloud is composed of particle masses between $10^{-6.5}$ and $10^{-3.5}$.[9] A broad spectrum of component percentages is given in the following table.

TABLE II

log m	-10.5		-9.5		-8.5		-7.5		-6.5		-5.5		-4.5		-3.5
%		0.1		1		3		8		16		21		18	
log m	-3.5		-2.5		-1.5		-0.5		+0.5		+1.5		+2.5		+3.5
%		14		9		5		3		1		0.5		0.2	

(After Millman)

Relating the density to the mass of the particle shows a general trend of increasing density with decreasing total mass. Table III lists estimates of these values.

TABLE III

Mass range (log m) (g)	Density ($g\ cm^{-3}$)
-1 to +1	0.3
-4 to -2	0.8
-14 to -8	3.0
<-14	>4. ?

(After Millman)

---

[9] *Ibid.*, p. 191.

The flux of the interplanetary dust seems to be a minimum between 0.7 and 0.8 AU with a gradual increase out to 3 AU where it is about twice the value at 1 AU. From there it begins to decrease again till at 5 AU its value is equal to the flux at 1 AU.[10] It is interesting to note that there is no dramatic increase near the asteroid belt as would be expected if it were a major source of dust.[11] The actual value of the flux at 1 AU is still an issue of debate due to disconcordant values calculated from data collected by diverse methods. The limitation of experiments to different size ranges is partially responsible for this. Also, the variations in the flux throughout the year make average values difficult to obtain. Even high altitude collection experiments have atmospheric contamination to contend with. Millman cites Barker and Anders who arrived at a flux of $4\times10^{-12}\ g\ m^{-2}\ s^{-1}$ from studies of deep-sea cores.[12]

Very little is known about physical structure besides densities. The major problem is in obtaining a dust particle which has not been altered in some way, either by impact with a collecting device or the earth's atmosphere. Some particles collected on upper air balloon and rocket flights have a fluffy type of structure with chains of small units held together by cohesive forces.[13] Meteorites which survive the fall to the earth are generally melted into globular, teardrop, or dumbbell shapes.

---

[10]*Ibid.*, p. 193.

[11]J. M. Alvarez, "The Cosmic Dust Environment at Earth, Jupiter and Interplanetary Space: Results from Langley Experiments on MTS, Pioneer 10, and Pioneer 11," in *Interplanetary Dust and Zodiacal Light*, eds. Hans Elsässer and H. Fechtig (New York, 1976), p. 181.

[12]Millman, p. 200.

[13]*Ibid.*, p. 197.

DUST ELIMINATION

The various processes which cause the dissipation of the interplanetary dust cloud are briefly reviewed here. Only solar gravitational attraction, solar radiation pressure and the Poynting-Robertson effect will be used in this thesis to calculate the lifetimes of particles in the solar system, but this survey will give an order of magnitude comparison with the other processes.

SPUTTERING

The solar wind is composed of atomic particles moving outward from the sun with a velocity of 600 km/s in the vicinity of the orbit of the earth. One effect of this wind on objects in the solar system is to cause an erosion of material from their surface. This is called sputtering and at 1 AU acts at an estimated rate of 0.5×10^{-7}mm/year.[14]

EVAPORATION

There is also an evaporation of the material of an interplanetary particle which increases dramatically as it approaches the sun due to the exponential dependence of the evaporation rate on the temperature. At 0.1 AU the rate may increase as high as 0.01 mm/year.[15] This rapid increase in evaporation has led to the suggestion that the inner zone of space around the sun is virtually dust free. Table IV gives a comparison of the rates of sputtering and evaporation as the sun is approached.

---

[14]S. Fred Singer and Lothar S. Bandermann, "Nature and Origin of Zodiacal Dust." in *The Zodiacal Light and the Interplanetary Medium*, ed. J. L. Weinberg (Washington, D.C., 1967), p. 394.

[15]*Ibid.*, pp. 393-394.

TABLE IV

COMPARISON OF SPUTTERING AND EVAPORATION RATES

Distance from Sun (AU)	Evaporation Rate for Stone (10^{-7}mm/yr)	Evaporation Rate for Iron (10^{-7}mm/yr)	Sputtering Rate Iron or Stone (10^{-7}mm/yr)
0.10	7.3×10^4	7.3×10^1	5.0
0.15	5.7×10^4	5.9×10^{-3}	2.2
0.20	7.4×10^{-2}	2.1×10^{-6}	1.3
0.25	2.1×10^{-4}	-	0.80
0.30	1.0×10^{-6}	-	0.56

(After Singer and Bandermann)

In general, the effect of sputtering in eliminating dust is negligible because of the rate with which the Poynting-Robertson effect causes infall and the associated rapid evaporation rate near the sun.

ELECTROMAGNETIC EFFECTS

There are three different types of electromagnetic forces which cause dissipation of solar system dust. Coulomb drag is an electrostatic interaction between the charge of the dust particle and the plasma ions and electrons. It is represented by

$$F_e = 10^{-23} n_e Z \frac{\lambda}{Au^2}$$

where Z is the number of elementary charges on the dust grain, A is the atomic weight of the plasma ions, u is the plasma velocity relative to the dust particle, n_e is the plasma density and λ is a number which depends on the charge. For typical conditions of interplanetary space $\lambda \approx 9$. There is also a Lorentz force on the particle due to interaction with the interplanetary field

$$F_L = \frac{Q}{c} \underset{\sim}{v} \times \underset{\sim}{B}$$

The convective force depends on the magnetic field carried by the solar wind with a high velocity, w, relative to that of the particle

$$F_{conv} = \frac{Q}{c} \underset{\sim}{w} \times \underset{\sim}{B}$$

Both the Coulomb drag and the Lorentz force are usually insignificant when compared with the Poynting-Robertson effect and the convective drag, which only becomes important for the particles with a radius <1 mm. Contrary to the Poynting-Robertson effect, the convective drag tends to increase the semimajor axis with decreasing eccentricity.[16]

PLANETARY PERTURBATIONS

The original studies of dust accretion by planets due to gravitational perturbations were done by Öpik. One of his most significant conclusions with respect to dust elimination was that particles larger than 1 mm will almost all be swept up by direct collisions with planets. The reason for this is the slow rate with which the Poynting-Robertson effect causes these particles to move in toward the sun. The particle moves so slowly through the path of the planets that the numerous passes of a planet will eventually guarantee its elimination from the dust cloud. It has also been shown that even a close encounter with a planet is likely to prevent the larger particles from drifting closer to the sun. Thus, the general action of the planets is that of a sieve by removing the

---

[16] *Ibid.*, pp. 395-396.

larger particles from the dust cloud.[17]

SELF COLLISIONS

The collision of asteroidal bodies is also considered a significant erosional process. The dust produced by the break up of these larger objects is then acted upon by dissipating effects such as radiation pressure and the Poynting-Robertson effect.

PSEUDO POYNTING-ROBERTSON EFFECT

The solar wind also produces an effect similar to that produced by solar radiation. The result of this effect is to increase the Poynting-Robertson effect on particles by 22 percent.[18]

RADIATION PRESSURE

Solar radiation produces a pressure on an interplanetary particle which depends on the cross sectional area of the particle and inversely on the square of the distance from the sun. The solar gravitational attraction can be exceeded by the radiation pressure for small particles due to the dependence of the force of gravity on the cube of the radius, and the dependence of radiation pressure on the square of the radius. The expression below gives the limiting value for the product of the density and radius of the particle

$$\rho a < 5.8\times10^{-5}\,\theta \qquad \text{c.g.s.}$$

where θ is the fraction of incident solar radiation effective in transfer-

---

[17] *Ibid.*, p. 396.

[18] Whipple, p. 411.

ring momentum to the particle.[19] Radiation pressure is quite effective in eradicating particles which originate in orbits of fairly high eccentricities. This is why long-period comets are not very useful in maintaining the dust cloud. For the radiation pressure to be most efficient the density of the particle must be fairly high. The value of θ depends on the electrical characteristics of the particle. For this reason the radiation is very effective in removing spherical iron particles from the solar system.[20]

ROTATIONAL BURSTING

A more recent area of study has been concerned with the rotation of a particle. Solar radiation pressure will produce a net torque on a particle which is not perfectly spherical. The resulting rotation may increase in frequency to the point at which the cohesive forces of the particle are overcome and it breaks apart. The theory indicates the possibility that this mechanism may be faster in eradicating particles than the Poynting-Robertson effect by as much as two orders of magnitude for nonmetallic particles and one order of magnitude for metallic particles.[21]

POYNTING-ROBERTSON EFFECT

The momentum of solar radiation produces a retarding force on a particle in orbit about the sun when the incident radiation is reradiated

---

[19] *Ibid.*, p. 410.

[20] Singer and Bandermann, p. 393.

[21] Stephen J. Paddack and John W. Rhee, "Rotational Bursting of Interplanetary Dust Particles," in *Interplanetary Dust and Zodiacal Light*, eds. H. Elsässer and H. Fechtig (New York, 1975), p. 453.

by the particle. The result of the drag force is to reduce the semimajor axis with a corresponding decrease of the eccentricity of the orbit. The combination of these changes in orbit causes the particle to spiral into the sun. The longest time of spiral in general is the case of a particle starting from a circular orbit. Öpik's study of planetary perturbations indicated an average lifetime of 10^8 years for particles with earth-crossing orbits and 10^6 for those with Jupiter-crossing orbits. For the Poynting-Robertson effect to be significant it must eliminate particles faster than this. For this reason it is limited in its major effectiveness to particles smaller than 10 cm in earth-crossing orbits and smaller than 0.1 cm for Jupiter-crossing orbits.[22]

DUST SOURCES

Whipple has summed up the state of the interplanetary dust cloud in the following quote:

> To date there is no evidence for any meteoritic matter with an immediate origin beyond the gravitational field of the sun, nor is there an indication of any accumulative processes on smaller bodies. The meteoritic complex is self-destructive.[23]

To maintain the dust cloud in a quasi-stable equilibrium, he has suggested a continuous input of 10 to 20 tons of dust per second would be necessary.[24] The two most plausible sources of dust will be considered in this section.

---

[22]Whipple, p. 411.

[23]*Ibid.*, p. 409.

[24]*Ibid.*, p. 420.

COMET DUST

The passage of a comet through the solar system leaves dust and ice particles in its path. Estimations of the total mass loss of a moderately bright comet whose perihelion distance is less than or equal to 1 AU are on the order of 10^{13} to 10^{15} grams.[25] As noted earlier, the radiation pressure can be very effective in removing cometary debris from the solar system. This factor must be evaluated in determining the possibility that comets could maintain the interplanetary dust cloud.

The average contributions of various comets are as follows. The short-period comets produce about 0.25 tons of dust per second. Over the last one hundred years the average production rate of long-period comets has been 4.9 tons/s. If the rare bright comets are included,a rate of 15.6 tons/s is obtained for all comets as bright as magnitude zero. When every comet is included an average of 20.8 tons/s is obtained. The actual value for the last century has been only 5 tons/s.[26]

If the value of 20.8 tons/s is taken as the proper value,it is seen that at least half of this must remain within the solar system. Even this may be unrealistic. It is suggested by some that "all but a very small fraction of the dust emitted is immediately lost by radiation pressure."[27]

---

[25]Vladimir Vanysek, "Dust in Comets and Interplanetary Matter," in *Interplanetary Dust and Zodiacal Light*, eds. H. Elsässer and H. Fechtig (New York, 1975), p. 229.

[26]A. H. Delsemme, "The Production Rate of Dust by Comets," in *Interplanetary Dust and Zodiacal Light*, eds. H. Elsässer and H. Fechtig (New York, 1975), pp. 315-317.

[27]S. Fred Singer, "Interplanetary Dust," in *Meteorite Research*, ed. Peter M. Millman (New York, 1969), p. 598.

That this is a fair estimate is apparent by the following two attempts to circumvent this problem. One suggestion is that a comet with a magnitude five times greater than Halley's Comet was captured into a short-period orbit about 20,000 years ago.[28] The other hypothesis is that the major source of dust is not the comet tail but unobservable cometary nuclei, left behind by the passage of comets, which provide a steady source of matter for meteoroid streams.[29]

ASTEROID DUST

It has already been observed that collisions in the asteroid belt will produce dust which will be added to the zodiacal cloud. Typical collisional models predict an increase in dust concentration in the asteroid belt of about one order of magnitude over that which exists at 1 AU.[30] Actual data from particle detection experiments on Pioneer 10 and 11 showed no increase in dust concentration in the region of the asteroid belt.[31] There is an obvious discrepancy between what was expected and what has been observed. It still remains to be explained completely. From the studies of the various models it can be concluded that the contribution of dust

---

[28] A. H. Delsemme, "Can Comets Be the Only Source of Interplanetary Dust?" in *Interplanetary Dust and Zodiacal Light*, eds. H. Elsässer and H. Fechtig (New York, 1975), p. 481.

[29] *Ibid.*, p. 483.

[30] J. S. Dohnanyi, "Sources of Interplanetary Dust: Asteroids," in *Interplanetary Dust and Zodiacal Light*, eds. H. Elsässer and H. Fechtig (New York, 1975), p. 189.

[31] J. M. Alvarez, "The Cosmic Dust Environment at Earth, Jupiter and Interplanetary Space: Results from Langley Experiments on MTS, Pioneer 10 and 11," in *Interplanetary Dust and Zodiacal Light*, eds. H. Elsässer and H. Fechtig (New York, 1975), p. 181.

from the asteroid belt to the interplanetary dust cloud is a small fraction of that contributed by comets.[32]

INTERSTELLAR DUST

As stated in the quote by Whipple, there does not seem to be any significant contribution of dust to the meteoritic complex from any source outside the solar system.

PART III

THE POYNTING-ROBERTSON EFFECT

HISTORICAL DEVELOPMENT

J. H. Poynting was the first to consider the effects of radiation on the orbits of small particles in the solar system. His application of classical physics in 1903 led him to the discovery that a particle which absorbs and reradiates solar radiation will be acted upon by a drag force. The result is a reduction of angular momentum of the particle causing it to spiral into the sun. His physical explanation for this drag was that it was due to a crowding of radiation in front of the particle with a corresponding thinning out behind. Poynting predicted that if the particle were small enough for gravitational attraction to be nearly balanced by radiation pressure this drag would produce a significant effect on the orbit of the particle.[33] It is most interesting to note that Poynting's resultant drag force derived with classical physics differed in magnitude from Robertson's by a ratio of only 2/3 to 1. Obviously, there has been no direct evidence to support either one over the other.

In 1913 J. Larmor developed an alternate treatment of the problem also based on classical electromagnetic theory.[34] He concluded that the

---

[33]J. H. Poynting, "Radiation in the Solar System: Its Effect on Temperature and Its Pressure on Small Bodies," *Philosophical Transactions of the Royal Society of London*, A, *202* (1903), p. 525-552.

[34]J. Larmor, *Proceedings of the 5th International Congress on Mathematics*, *1* (Cambridge, 1913), p. 197ff.

retarding force was solely due to the radiation of the particle and obtained a value for the retarding force that was three times greater than Poynting's. At that time Larmor was unaware that his result contradicted the theory of relativity which does not allow the reradiation alone to be the braking force on the particle. The above contradiction was demonstrated by L. Page in 1918 when he reworked the problem himself.[35] Larmor acknowledged this fact and then gave the correct relativistic explanation of the drag in a postscript to Poynting's *Collected Scientific Papers*.

> But for Poynting's particle describing a planetary orbit the radiation from the Sun comes in, which restores the energy lost by radiation from the particle, and so establishes again the retarding force $-(1/c^2)(dE/dt)v$.[36]

The most complete development applying the special theory of relativity was performed by H. P. Robertson in 1937. His equations of motion are now generally accepted as correct for a particle which absorbs all incident radiation and reradiates it isotropically in the frame of the particle. Part of the key to his success came from the expression of his results in the frame of reference of the sun, rather than attempting to study the retarding force in the frame of the particle as had been done in the past.[37]

In 1950 Wyatt and Whipple published a paper in which they applied

---

[35]L. Page, *Physical Review, 11* (1918), p. 376ff; *Physical Review, 12* (1918), p. 371ff.

[36]J. Larmor, "Postscript," *Collected Scientific Papers*, ed. Gilbert A. Shakespear, (Cambridge, 1920), p. 756.

[37]H. P. Robertson, "Dynamical Effects of Radiation in the Solar System," *Monthly Notices of the Royal Astronomical Society, 97* (April, 1937), pp. 423-438.

Robertson's equations to the calculation of the expected lifetime of a particle. They also predicted grading of meteor showers with the smaller particles closer to the sun. Their observations of the changes in magnitude of meteor showers did not give any definitive results except that the showers are probably fairly young.[38]

A. E. Guess published a paper in 1962 in which he studied the change of the drag force as a particle approaches the sun. For this case the sun can no longer be considered a point source of radiation but must be a spherical source of finite dimensions. His derivations showed an appreciable increase in the Poynting-Robertson effect near the sun.[39]

Recently R. A. Lyttleton has completed a more realistic study of the problem by assuming that there is some reflection of the incident radiation. The net result of this is to increase the repulsion effect of the radiation and reduce the amount of the drag force on the particle.[40]

EQUATIONS OF MOTION

Due to the general acceptance of the work done by Robertson, his equations of motion are now simply assumed as correct in most treatments of this problem. This section will deal with the derivation of these equations in minute detail including as much physical discussion as deemed necessary for a straightforward comprehension of what has been done. The

---

[38]Stanley P. Wyatt, Jr. and Fred L. Whipple, "The Poynting-Robertson Effect on Meteor Orbits," *Astrophysical Journal, 111* (1950), pp. 134-141.

[39]A. E. Guess, "Poynting-Robertson Effect for a Spherical Source of Radiation," *Astrophysical Journal, 135* (1962), pp. 855-866.

[40]R. A. Lyttleton, "Effects of Solar Radiation on the Orbits of Small Particles," *Astrophysics and Space Science, 44* (1976), pp. 119-140.

essence of what Robertson has done will be maintained in order to make this work directly comparable to his paper.

The physical phenomenon of interest can be described in the following way. A small spherical body is moving in an inertial system, S, whose origin coincides with the center of the sun. All incident radiation is absorbed and then reradiated isotropically in the reference system whose origin moves with the body. Rotational motion of the body is ignored.

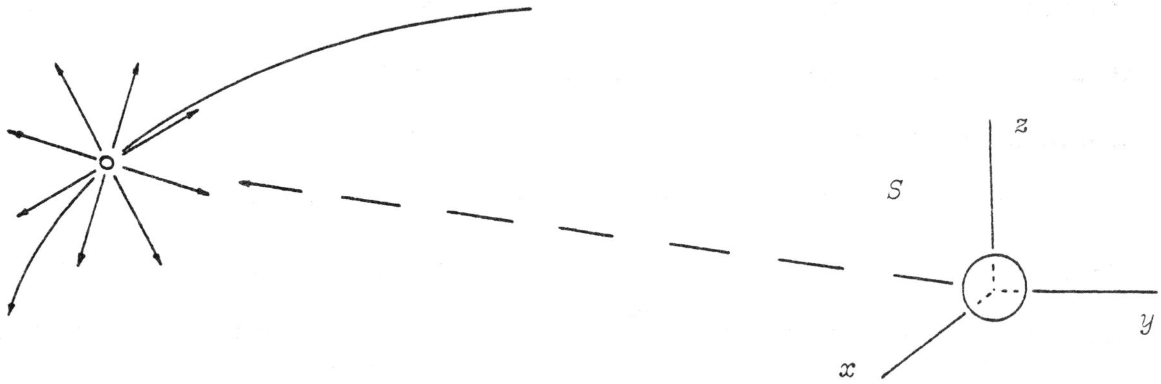

Figure (2). Particle reradiating energy from sun.

Since this derivation employs the theory of special relativity it is convenient to use a Minkowski coordinate system of the form $x^{\mu} = (t,x,y,z)$ where the range of μ is (0,1,2,3). A world line element in the inertial system of the sun is

$$ds^2 = g_{\mu\nu} dx^{\mu} dx^{\nu} = dt^2 - (dx^2 + dy^2 + dz^2)/c^2 \qquad (1)$$

where ν obviously has the same range as μ.[41]

[41]The notation employed in this thesis is of the form described by A. S. Eddington in his book *The Mathematical Theory of Relativity*, (Cambridge, 1957). Appendix II at the end of this work describes the basic operations necessary for an understanding of the development of the equations of motion using the special theory of relativity.

A unit four-vector, u^μ, is chosen to represent the velocity of the particle such that

$$g_{\mu\nu}u^\mu u^\nu = 1 \ . \tag{2}$$

Simply stated, its magnitude in four-space is equal to one. The energy-momentum vector is written as mu^μ, where m is the rest mass of the body.

The incident radiation is assumed to arrive at the particle in plane-parallel waves, implying that the source of radiation is at a great distance with respect to the dimensions of the particle. Thus, for these calculations it is proper to consider the sun as a point source of radiation. The direction of this radiation is defined by a null-vector, l^μ, with the normalization condition $l^0 = 1$. This is expressed in tensor notation as

$$g_{\mu\nu}l^\mu l^\nu = 0 \qquad \text{where } l^0 \equiv 1 \ . \tag{3}$$

The energy density for the incident radiation can be expressed as a constant time-average value, d. The relation of this magnitude and the beam

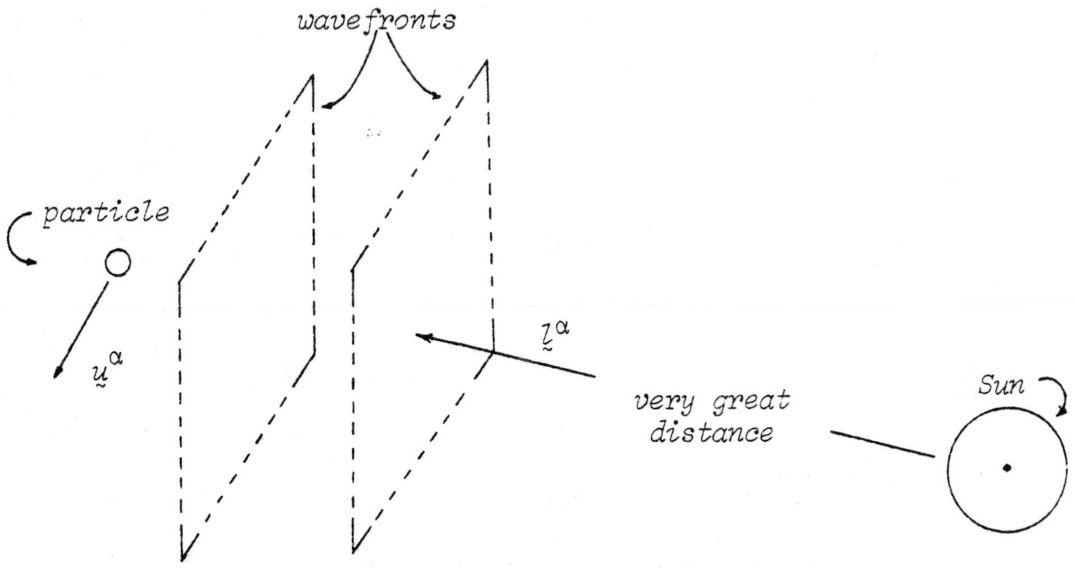

Figure (3). Plane-parallel waves from distant source.

direction is expressed by the following tensor,

$$e_{\mu\nu} = dl^{\mu}l^{\nu} \ . \tag{4}$$

Clearly, for the case $\mu,\nu = 0$, $e^{00} = d$ since $l_0^2 = 1$. This is the value of momentum of radiation carried per second across a unit area oriented perpendicular to the direction l^{α} where $\alpha = (1,2,3)$. This parameter d is also equivalent to a pressure acting in the spatial direction l^{α}.

As stated earlier, one of the major differences between the approach of Robertson and earlier derivations was that he obtained the equations of motion in the frame of the particle and then transformed those expressions into the rest system of the sun. Following this method, it is then necessary to express the above values in terms of a frame of reference whose origin at some instant, E, coincides with the position of the particle. Let this system be designated as Σ_E with Minkowski coordinates $\xi^{\mu} = (\tau,\xi,\eta,\zeta)$. The velocity of the particle, u^{μ}, and the direction of the radiation, l^{μ}, are defined in Σ_E by υ^{μ} and λ^{μ} respectively. Then, at E

$$\upsilon^{\mu} = \delta_0^{\mu} \ . \tag{5}$$

This expression implies that the instantaneous velocity of the body is zero in the spatial dimensions of Σ_E at E. Note that the four-vector velocity in Σ_E is a unit vector since $\upsilon^0 = 1$. In order to express the direction of the radiation in Σ_E the freedom does not exist to define $\lambda^0 = 1$ as was done for l^0

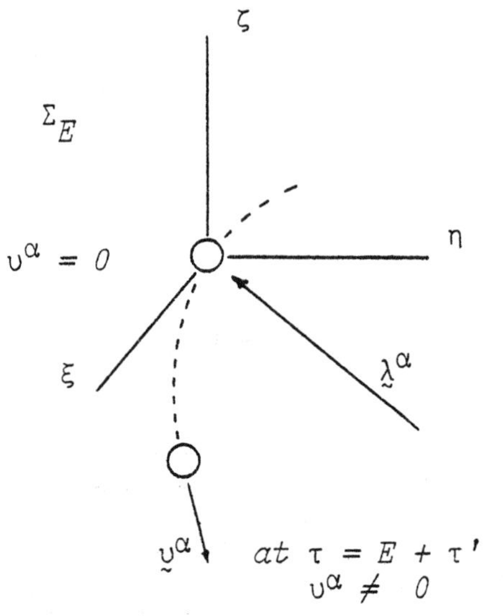

Figure (4). Radiation incident on particle in Σ_E at E.

because the value of λ^0 is derived from l^μ and is restricted to the result of this derivation. Since λ^μ is a null-vector, which will be true for any four-vector representing electromagnetic radiation, and since $\lambda^0 \neq 1$, the values of the spatial components may be obtained in the following way.

$$g_{\mu\nu}\lambda^\mu\lambda^\nu = \lambda_0^2 - (\lambda_1^2 + \lambda_2^2 + \lambda_3^2)/c^2 = 0 \qquad (6)$$

For this to be true the expression in parentheses must equal $\lambda_0^2 c^2$. Let

$$\lambda^\alpha = c\lambda^0\nu^\alpha \qquad \text{where } \alpha = (1,2,3) \; . \qquad (7)$$

Then substitution in (6) gives

$$\lambda_0^2 - \left[(c\lambda_0\nu_1)^2 + (c\lambda_0\nu_2)^2 + (c\lambda_0\nu_3)^2\right]/c^2 = 0$$

$$\lambda_0^2 - c^2\lambda_0^2(\nu_1^2 + \nu_2^2 + \nu_3^2)/c^2 = 0 \qquad (8)$$

For this to be true the expression in the parentheses in (8) must equal one. Thus, ν^α represents a unit vector which defines the spatial direction of the radiation in Σ_E. With this interpretation (7) is a proper expression for the spatial components of λ^α in Σ_E. The energy-momentum stress tensor due to solar radiation is

$$\varepsilon^{\mu\nu} = \frac{\delta\lambda^\mu\lambda^\nu}{\lambda_0^2} \qquad (9)$$

where λ_0^2 is the normalization factor. This provides the result that $\varepsilon^{00} = \delta$ which is the energy density of the radiation in Σ_E.

Since the particle is reradiating the energy it receives from the sun as quickly as it absorbs it there will be no net gain in the energy of the particle which implies that the rest mass is constant. With this and the fact that υ^μ is a unit vector it is clear that the time component of the momentum of the particle at E is a constant, *i.e.*

$$d(m\upsilon^0)/d\tau = 0 \; . \qquad (10)$$

The spatial component of the momentum $m\upsilon^{\alpha}$ changes at a rate which is equal to the force acting on the particle even though the instantaneous value of $m\upsilon^{\alpha}$ at E is zero. The direction of the force is the same as ν^{α} since the particle is spherical and the radiation from the particle is isotropic in Σ_E. Classical physics can be used to determine the magnitude of this force since the particle is instantaneously at rest in Σ_E. The energy density, δ, of the incoming beam of radiation acts as a pressure on the cross-sectional area of the particle. The force, ϕ, acting on the body is given by

$$\phi = A\delta \tag{11}$$

where A can be interpreted as being equal to πa^2 and a represents the radius of the particle for the case where it is large with respect to the wavelength of the radiation. When the size of the particle is small enough for diffraction effects to become significant, Robertson has designated A as the "effective cross-section" which can be approximated from theoretical calculations.

The equations of motion will simply be an expression of Newton's second law in the system Σ_E. The derivative of the momentum with respect to time is

$$\frac{d(m\upsilon^{\mu})}{d\tau} . \tag{12}$$

The force applied by the incident radiation was defined above as ϕ and is in the direction ν^{α}. By (7), if λ^{μ} is divided by $c\lambda^0$ it will be normalized. Then the proper expression for the total power-force four-vector due to the incident radiation, in terms of λ^{μ}, will be

$$\frac{\phi\lambda^{\mu}}{c\lambda^0} . \tag{13}$$

The effect of the reradiated energy can be explained physically in the following manner. The radiation of energy is the equivalent of a loss of mass which depends on the rate at which the energy is radiated. This implies that there will be a loss of particle momentum with time which further implies that there is a net force exerted on the particle. Since this force is reducing the momentum of the particle it must be acting in the direction opposite that of the velocity. The rate at which energy is radiated is simply ϕc. If Δe is the energy loss per second, the mass loss per second, Δm, equals $\Delta e/c^2$ since $\Delta e = \Delta mc^2$. The change in the momentum due to a loss of mass is then

$$-\frac{\phi c}{c^2}\upsilon^\mu \quad \text{or} \quad -\frac{\phi}{c}\upsilon^\mu \,. \tag{14}$$

Setting (12) equal to (13) and (14) gives the total change in momentum

$$\frac{d(m\upsilon^\mu)}{d\tau} = \frac{\phi\lambda^\mu}{c\lambda^0} - \frac{\phi}{c}\upsilon^\mu$$

which is written more concisely as

$$\frac{d(m\upsilon^\mu)}{d\tau} = \frac{\phi}{c\lambda^0}(\lambda^\mu - \lambda^0\upsilon^\mu) \,. \tag{15}$$

The conditions noted earlier still hold, *i.e.* the change of the time component of the momentum is zero.

$$\frac{d(m\upsilon^0)}{d\tau} = \frac{\phi}{c\lambda^0}(\lambda^0 - \lambda^0\upsilon^0)$$

Since $\upsilon^0 = 1$

$$\frac{d(m\upsilon^0)}{d\tau} = \frac{\phi}{c\lambda^0}(\lambda^0 - \lambda^0) = 0 \,. \tag{16}$$

The spatial component where $\mu = \alpha$ is of the form

$$\frac{d(mv^{\alpha})}{d\tau} = \frac{\phi}{c\lambda^0}(\lambda^{\alpha} - \lambda^0 v^{\alpha})$$

$$\frac{d(mv^{\alpha})}{d\tau} = \frac{\phi\lambda^{\alpha}}{c\lambda^0} - \frac{\phi}{c} v^{\alpha} \, . \tag{17}$$

At E (17) is

$$\frac{d(mv^{\alpha})}{d\tau} = \frac{\phi\lambda^{\alpha}}{c\lambda^0}$$

since v^{α} equals zero at E. The only change in momentum is due to the applied force of radiation because the reradiation produces no effect on the particle momentarily at rest in Σ_E.

Equation (15) is the equation of motion in Σ_E. This must now be transformed into an equation which is defined with respect to the rest system S with coordinates x^{μ} as defined earlier. The values λ^{μ} and v^{μ} are simply the components of the four-vectors in Σ_E which express the direction of the identical radiation beam and the motion of the same particle as expressed by l^{μ} and u^{μ} respectively in S. For this reason they can be interchanged without any conversion factor. The scalar quantity λ^0 in Σ_E is the dot product of λ^{μ} with v^{μ} and can be represented in S as

$$w = l_{\mu}u^{\mu} = \lambda^0 \tag{18}$$

by reason of the statement immediately above. The radiation force, ϕ, on the particle depends on δ which is the component ε^{00} of the energy-momentum stress tensor in Σ_E. The value of δ may also be represented as the double inner product $\varepsilon^{\mu\nu}v_{\mu}v_{\nu}$ which in S is $e^{\mu\nu}u_{\mu}u_{\nu}$. But

$$e^{\mu\nu}u_{\mu}u_{\nu} = dl^{\mu}l^{\nu}u_{\mu}u_{\nu}$$

by (4), and

$$dl^{\mu}l^{\nu}u_{\mu}u_{\nu} = dl^{\mu}u_{\mu}l^{\nu}u_{\nu}$$

$$= dw^2$$

by (18). Thus

$$\delta = w^2 d \tag{19}$$

and

$$\phi = w^2 f \tag{20}$$

where $f = Ad$ and represents the force which would act on the particle if it were at rest in S.

With the above information the equations of motion in S can be written as

$$\frac{dmu^{\mu}}{ds} = \frac{fw}{c}(l^{\mu} - wu^{\mu}) \tag{21}$$

where s is the proper time of the particle.

A brief look at the implications of these equations is in order at this point. Mass must be conserved and this can be shown by multiplying through by u_{μ}.

$$u_{\mu}\left(\frac{dmu^{\mu}}{ds}\right) = u_{\mu}\frac{fw}{c}(l^{\mu} - wu^{\mu})$$

$$\frac{dmu_{\mu}u^{\mu}}{ds} = \frac{fw}{c}(u_{\mu}l^{\mu} - wu_{\mu}u^{\mu}) .$$

By (2), when $u_{\mu}u^{\mu}$ is summed over μ it equals one, and by (18) $u_{\mu}l^{\mu}$ equals w. Therefore

$$\frac{dm}{ds} = \frac{fw}{c}(w - w) = 0 . \tag{22}$$

The validity of Larmor's statement concerning the existence of the retarding force can also be demonstrated at this time. The second term on the right side of (21) is obviously an actual drag force, but its presence must be associated with the energy absorbed from the incident beam of radiation. If not, there will be no reduction of the velocity of the particle. This is clearly seen if the term fwl^{μ}/c, representing the effect of the incident energy, is removed from (21) leaving

$$\frac{dmu^{\mu}}{ds} = -\frac{fw^2}{c}u^{\mu} .$$

Then

$$m\frac{du^{\mu}}{ds} + \frac{dm}{ds}u^{\mu} = -\frac{\phi}{c}u^{\mu} \tag{23}$$

by (20). The term on the right can be analyzed dimensionally using the fact that ϕc is the rate at which energy is being radiated. Thus,

$$\frac{\phi}{c} = \frac{\phi c}{c^2} = \left(\frac{erg/s}{cm^2/s^2}\right) = \left(\frac{g\text{-}cm^2/s^2}{cm^2/s}\right) = \left(\frac{g}{s}\right) ,$$

and the term on the right represents the rate at which mass is being lost by the particle. The second term on the left is then obviously canceled out by the right side of (23) leaving

$$m\frac{du^{\mu}}{ds} = 0$$

and the conclusion that the velocity will remain constant. Therefore, unless the beam of radiation continues to supply energy to the particle, it will not experience a retarding force which is what Larmor concluded.

Since there are no significant relativistic effects involved, the following Newtonian approximations will be permissible and convenient to make. The guideline for approximations will be to retain only terms of the first order in the ratio of particle velocity to the speed of light. The vectors representing the velocity of the particle and the direction of the incident beam can be written as

$$\mathrm{v} = \frac{dx^\alpha}{dt} \underset{\sim}{l}_\alpha \quad \text{and} \quad \mathrm{n} = \frac{l^\alpha}{c} \underset{\sim}{l}_\alpha \quad \text{where } \alpha = (1,2,3) \tag{24}$$

and $\mathbf{n}$ is obviously a unit vector. Applying the guideline for approximations allows the other terms in the equations of motion to be written as

$$u^0 = 1\ , \quad \underset{\sim}{u} = \mathrm{v}\ , \quad \text{and} \quad w = 1 - \mathrm{v}\cdot\mathrm{n}/c\ . \tag{25}$$

Substituting these expressions in (21) gives

$$m\frac{d\mathrm{v}}{dt} = \frac{f}{c}(1 - \mathrm{v}\cdot\mathrm{n}/c)\big(c\,\mathrm{n} - (1 - \mathrm{v}\cdot\mathrm{n}/c)\,\mathrm{v}\big)$$

$$= f(1 - \mathrm{v}\cdot\mathrm{n}/c)\,\mathrm{n} - \frac{f}{c}(1 - \mathrm{v}\cdot\mathrm{n}/c)^2\,\mathrm{v}$$

$$= f(1 - \mathrm{v}\cdot\mathrm{n}/c)\,\mathrm{n} - \frac{f}{c}\big(1 - 2\mathrm{v}\cdot\mathrm{n}/c + (\mathrm{v}\cdot\mathrm{n})^2/c^2\big)\,\mathrm{v}$$

$$= f(1 - \mathrm{v}\cdot\mathrm{n}/c)\,\mathrm{n} - \frac{f}{c}\mathrm{v} + \frac{2f\mathrm{v}\cdot\mathrm{n}}{c^2}\mathrm{v} - \frac{f(\mathrm{v}\cdot\mathrm{n})^2}{c^3}\mathrm{v}\ .$$

By the guideline for approximations,

$$m\frac{d\mathrm{v}}{dt} = f(1 - \mathrm{v}\cdot\mathrm{n}/c)\,\mathrm{n} - \frac{f}{c}\mathrm{v}\ . \tag{26}$$

The first term on the right is the radiation pressure in the direction $\mathbf{n}$ reduced due to the Doppler effect by $\mathrm{v}\cdot\mathrm{n}/c$. The second term is the drag force. If the velocity of the particle is resolved into its components parallel and perpendicular to the beam direction $\mathbf{n}$, (26) can be rewritten in the form

$$m \frac{d\mathbf{v}}{dt} = f(1 - \mathbf{v}\cdot\mathbf{n}/c)\,\mathbf{n} - \frac{f\mathbf{v}\cdot\mathbf{n}}{c}\,\mathbf{n} - \frac{f}{c}\,\mathbf{v}'$$

where $\mathbf{v}'$ is the vector component of $\mathbf{v}$ perpendicular to $\mathbf{n}$. Thus

$$m \frac{d\mathbf{v}}{dt} = f(1 - 2\mathbf{v}\cdot\mathbf{n}/c)\,\mathbf{n} - \frac{f}{c}\,\mathbf{v}' . \tag{27}$$

As described earlier, the particle of interest is moving in an orbit about the sun with complete absorption and reradiation of the solar energy incident upon it. This radiation is propagating radially outward from the sun which by definition is the direction $\mathbf{n}$. It also decreases in its intensity as the inverse of the square of the solar distance. Let the symbol § be defined to represent the solar constant which is the amount of energy per second incident on one square centimeter at a distance of 1 AU. This value has been measured to be equal to 1.36×10^{6} $erg\ s^{-1}\ cm^{-2}$. Thus the energy density at some distance r is expressed as

$$d = \frac{\S b^2}{cr^2} \tag{28}$$

where b equals 1 AU and the force due to the radiation is

$$f = Ad = \frac{A\S b^2}{cr^2} . \tag{29}$$

For convenience in later calculations let

$$\alpha \equiv \frac{A\S b^2}{mc^2} = \frac{\pi a^2 \S b^2}{(4/3)\pi a^3 \rho c^2} = \frac{3\S b^2}{4a\rho c^2} \tag{30}$$

where a is the radius of the particle and ρ is its density. Then

$$f = \frac{A\S b^2}{cr^2} = \left(\frac{A\S b^2}{mc^2}\right)\frac{mc}{r^2} = \frac{\alpha mc}{r^2} . \tag{31}$$

The value of α can be calculated using the above value for §,

$b = 1.496\times10^{13}$ *cm*, and $c = 2.998\times10^{10}$ *cm/s*.

$$\alpha = \frac{2.54\times10^{11}}{a\rho}\ cm^2/s \qquad (32)$$

This applies for the cases where diffraction effects are negligible. To complete the description of the forces acting on the particle the force of solar gravitational attraction must be included. It can be expressed as GMm/r^2 where G is the gravitational constant and M is the mass of the sun. It is acting in the $-\mathbf{n}$ direction. The above expressions can now be used to rewrite (27) as

$$m\frac{d\mathbf{v}}{dt} = \left(\frac{\alpha mc}{r^2}(1 - 2\mathbf{v}\cdot\mathbf{n}/c) - \frac{GMm}{r^2}\right)\mathbf{n} - \frac{\alpha m}{r^2}\mathbf{v}' \ . \qquad (33)$$

From the form of the equation above it is clear that it will be most convenient to express the equations of motion in polar coordinates (r,θ). For the simple case of a mass traveling about a center of force, the polar equations of motion can be derived in the following way.

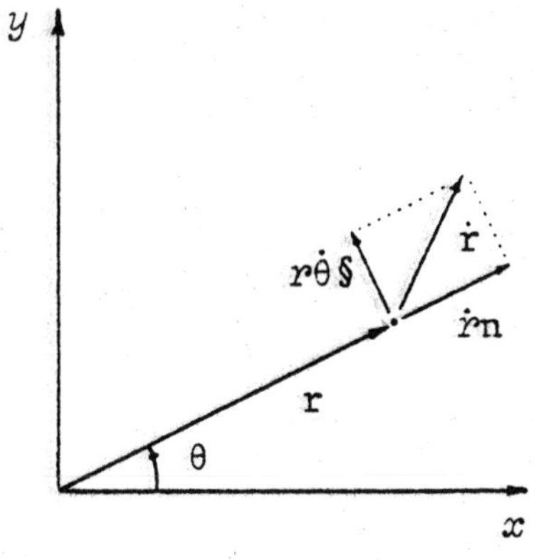

Figure (5).

If the position of the mass is defined by $\mathbf{r}$, the change in position with time is represented in figure (5) where $\mathbf{n}$ and § are the unit vectors in the r and θ directions respectively. Then

$$\mathbf{n} = \cos\theta\ \underset{\sim}{i} + \sin\theta\ \underset{\sim}{j} \qquad (34a)$$

and

$$§ = -\sin\theta\ \underset{\sim}{i} + \cos\theta\ \underset{\sim}{j} \qquad (34b)$$

where $\underset{\sim}{i}$ and $\underset{\sim}{j}$ are unit vectors in the x and y directions. Then

$$\frac{d\mathbf{n}}{dt} = -\frac{d\theta}{dt}\sin\theta\ \underset{\sim}{i} + \frac{d\theta}{dt}\cos\theta\ \underset{\sim}{j}$$

$$\frac{dn}{dt} = \frac{d\theta}{dt}(-\sin\theta\ \underset{\sim}{i} + \cos\theta\ \underset{\sim}{j})$$

$$\frac{dn}{dt} = \frac{d\theta}{dt}\ \S\ . \qquad (35)$$

And

$$\frac{d\S}{dt} = -\frac{d\theta}{dt}\cos\theta\ \underset{\sim}{i} - \frac{d\theta}{dt}\sin\theta\ \underset{\sim}{j}$$

$$= -\frac{d\theta}{dt}(\cos\theta\ \underset{\sim}{i} + \sin\theta\ \underset{\sim}{j})$$

$$\frac{d\S}{dt} = -\frac{d\theta}{dt}\ n\ . \qquad (36)$$

The first derivative of r with respect to time is

$$\frac{d\mathbf{r}}{dt} = \frac{dr}{dt}\ n + r\frac{d\theta}{dt}\ \S\ . \qquad (37)$$

The second derivative is expressed as

$$\frac{d^2\mathbf{r}}{dt^2} = \frac{d^2r}{dt^2}\ n + \frac{dr}{dt}\frac{dn}{dt} + \frac{dr}{dt}\frac{d\theta}{dt}\ \S + r\frac{d^2\theta}{dt^2}\ \S + r\frac{d\theta}{dt}\frac{d\S}{dt}$$

$$= \frac{d^2r}{dt^2}\ n + \frac{dr}{dt}\frac{d\theta}{dt}\ \S + \frac{dr}{dt}\frac{d\theta}{dt}\ \S + r\frac{d^2\theta}{dt^2}\ \S - r\left(\frac{d\theta}{dt}\right)^2 n$$

by (35) and (36). Collecting radial and angular components

$$\frac{d^2\mathbf{r}}{dt^2} = \left[\frac{d^2r}{dt^2} - r\left(\frac{d\theta}{dt}\right)^2\right] n + \left[2\frac{dr}{dt}\frac{d\theta}{dt} + r\frac{d^2\theta}{dt^2}\right] \S\ . \qquad (38)$$

Since

$$\frac{1}{r}\frac{d}{dt}\left(r^2\frac{d\theta}{dt}\right) = \frac{1}{r}\left(2r\frac{dr}{dt}\frac{d\theta}{dt} + r^2\frac{d^2\theta}{dt^2}\right) = 2\frac{dr}{dt}\frac{d\theta}{dt} + r\frac{d^2\theta}{dt^2}\ ,$$

the angular acceleration term in (38) can be replaced with

$$\frac{1}{r}\frac{d}{dt}\left(r^2\frac{d\theta}{dt}\right).$$

Equating the product of the particle mass and the radial acceleration to the component of the force in (33) in the $\mathbf{n}$ direction gives

$$m\left[\frac{d^2r}{dt^2} - r\left(\frac{d\theta}{dt}\right)^2\right] = \frac{\alpha mc}{r^2}\left(1 - \frac{2}{c}\frac{dr}{dt}\right) - \frac{GMm}{r^2}.$$

The mass drops out leaving

$$\frac{d^2r}{dt^2} - r\left(\frac{d\theta}{dt}\right)^2 = \frac{\alpha c}{r^2} - \frac{GM}{r^2} - \frac{2\alpha}{r^2}\frac{dr}{dt}$$

$$\frac{d^2r}{dt^2} - r\left(\frac{d\theta}{dt}\right)^2 = -\frac{\mu}{r^2} - \frac{2\alpha}{r^2}\frac{dr}{dt} \qquad (39)$$

where for convenience $GM - \alpha c$ is defined as μ and represents the net effect of gravitational attraction and radiation pressure. Also, the product of the mass and the angular acceleration equated to the transverse component of the force in (33) gives

$$m\frac{1}{r}\frac{d}{dt}\left(r^2\frac{d\theta}{dt}\right) = -\frac{\alpha mr}{r^2}\frac{d\theta}{dt}$$

or

$$\frac{1}{r}\frac{d}{dt}\left(r^2\frac{d\theta}{dt}\right) = -\frac{\alpha}{r}\frac{d\theta}{dt}. \qquad (40)$$

The angular momentum per unit mass, h, is being reduced at a rate which can be determined by integrating (40).

$$d\left(r^2\frac{d\theta}{dt}\right) = -\alpha\, d\theta$$

By simple integration

$$h = r^2 \frac{d\theta}{dt} = - \alpha\theta + H \qquad (41)$$

where H is defined as the initial angular momentum. Total loss of angular momentum will result in a direct fall into the sun or explusion from the solar system by radiation pressure. To find out how many revolutions can take place before this occurs, let $h = 0$. Then $\theta = H/\alpha$ or

$$R = \frac{H}{2\pi\alpha} \qquad (42)$$

where R represents the number of revolutions possible. The value of R only depends on the initial angular momentum, the size and the density of the particle.

EFFECT OF PARTIAL REFLECTION

A modification of the equations of motion was proposed by R. A. Lyttleton to take into account the effect of reflection of some fraction of the incident light on the particle. This is certainly a more realistic statement of the problem since only a perfect black body could comply with the assumptions made by Robertson. Since there is little known directly concerning the physical properties of an interplanetary particle in this respect, it is only possible to motivate the modified form of the equations of motion by discussing the most probable way the particle will interact with the incident radiation.

As already stated, there will certainly be some reflection of solar radiation from the surface of a particle. Also, some rotation of the particle would be expected since it is highly unlikely that it would be perfectly spherical. The radiation pressure will produce some net torque resulting in a rotation. It would be excessively cumbersome to express this rotation as a general function of the shape of the particle, and even if this were done there would be no way to associate this directly with reality except in a very general way. For these reasons it will simply be assumed that the time average effect of this reflection is defined in direction by a unit vector **k** associated with the average direction of the reflection during a short interval of time.

A modification must be made to express the energy now involved in the Poynting-Robertson drag. If Ψ is the total radiation energy per second incident on the particle, let Ψ_i be the part absorbed and reradiated isotropically so that the portion reflected in the direction **k** is $\Psi - \Psi_i$. The equivalent physical situation can be described as the energy Ψ being incident along **n** and reradiated isotropically while $-(\Psi - \Psi_i)$ is incident along **k** and reradiated isotropically. Equation (26) can be rewritten

$$m \frac{d\mathbf{v}}{dt} = f(1 - \mathbf{v \cdot n}/c)\,\mathbf{n} - \frac{\Psi_i}{\Psi}\frac{f}{c}\,\mathbf{v} - \frac{\Psi - \Psi_i}{\Psi} f(1 - \mathbf{v \cdot k}/c)\,\mathbf{k} \qquad (44)$$

where the drag force is now due to Ψ_i rather than the total amount of the radiation Ψ. For convenience let $(\Psi - \Psi_i)/\Psi$ be represented by ψ, and the above equation becomes

$$m \frac{d\mathbf{v}}{dt} = f(1 - \mathbf{v \cdot n}/c)\,\mathbf{n} - (1 - \psi)\frac{f}{c}\,\mathbf{v} - \psi f(1 - \mathbf{v \cdot k}/c)\,\mathbf{k}\ . \qquad (45)$$

The orientation of **k** with respect to $-$ **n** can be described by the polar angle λ and the azimuthal angle χ such that $\mathbf{k \cdot n} = -\cos\lambda$. Figure (6) shows the arrangement of these angles. With regard to the actual physical situation, it would be expected that the largest portion of the

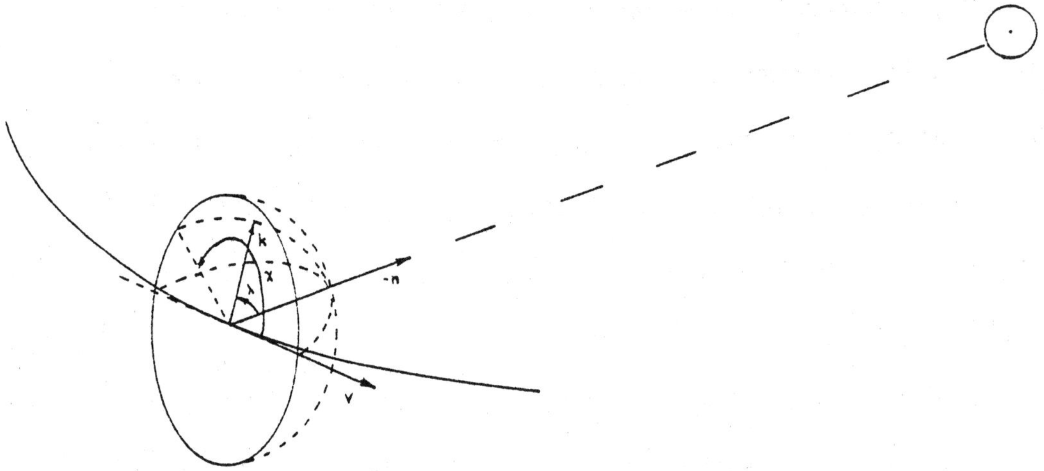

Figure (6). Particle reflecting light in **k** direction.

radiation would be reflected back toward the sun. This implies that **k** will always lie in the hemisphere toward the sun and most likely make a small angle with **- n**. In order to conform with this analysis of the reflection, Lyttleton has suggested an empirical mathematical expression of the form

$$\frac{\kappa + 1}{2\pi} \cos^{\kappa}\lambda \qquad (46)$$

which represents the probability per unit area of the sphere that **k** will be pointing in that direction.[42] The $\cos^{\kappa}\lambda$ is the initial expression, but the total probability over all the hemisphere must equal one. An element of area of the hemisphere is $\sin\lambda d\lambda d\chi$ so that if $\cos^{\kappa}\lambda$ is integrated over the hemisphere the result is

$$\int_0^{2\pi}\int_0^{\pi/2} \cos^{\kappa}\lambda \sin\lambda d\lambda d\chi$$

---

[42]Personal communication with Dr. R. A. Lyttleton, April 12, 1978.

which by integrating with respect to χ is

$$2\pi\int_0^{\pi/2} \cos^{\kappa}\lambda \sin\lambda \, d\lambda$$

and with respect to λ is

$$2\pi\left(-\frac{\cos^{\kappa+1}\lambda}{\kappa + 1}\right)_0^{\pi/2} = \frac{2\pi}{\kappa + 1} .$$

Since the total probability should be one, $\cos^{\kappa}\lambda$ is normalized by dividing by $2\pi/(\kappa + 1)$ which gives expression (46). The symbol κ is not specifically defined except in the sense that it determines how sharp the peak of the probability will be as $\lambda \to 0$. A physical interpretation of the effect of κ is that its magnitude determines the concentration of the reflected radiation back in the direction of the sun.

In order to arrive at a time average expression for the force acting on the particle it will be helpful to express k and v in their radial and angular components. Thus

$$\mathbf{k} = - \cos\lambda \; \mathbf{n} + \sin\lambda\cos\chi \; \S \tag{47}$$

and

$$\mathbf{v} = \frac{dr}{dt}\,\mathbf{n} + r\,\frac{d\theta}{dt}\,\S \; . \tag{48}$$

The following dot products are evaluated as

$$\mathbf{v}\cdot\mathbf{n} = \frac{dr}{dt}$$

$$\mathbf{v}\cdot\mathbf{k} = - \cos\lambda \frac{dr}{dt} + \sin\lambda\cos\chi\left(r\,\frac{d\theta}{dt}\right) .$$

These two expressions can now be substituted in (45) to give a form of the equation with only unit vectors and scalor quantities.

$$m\frac{dv}{dt} = f\left(1 - \frac{1}{c}\frac{dr}{dt}\right)\mathrm{n} - (1 - \psi)\frac{f}{c}\frac{dr}{dt}\mathrm{n} - (1 - \psi)\frac{f}{c}\left(r\frac{d\theta}{dt}\right)\S$$

$$- \psi f\left\{1 - \frac{1}{c}\left[- cos\lambda\frac{dr}{dt} + sin\lambda cos\chi\left(r\frac{d\theta}{dt}\right)\right]\right\}(- cos\lambda)\ \mathrm{n}$$

$$- \psi f\left\{1 - \frac{1}{c}\left[- cos\lambda\frac{dr}{dt} + sin\lambda cos\chi\left(r\frac{d\theta}{dt}\right)\right]\right\}(sin\lambda cos\chi)\ \S\ . \quad (49)$$

Multiplying the probability expression through (49) and then integrating over the hemisphere toward the sun will give the time average value of this equation of motion. Since the probability function is normalized, those terms not dependent on λ and χ will just be multiplied by one. Therefore, in the following averaging procedure only the last two terms of (49) will be dealt with. Multiplying these terms out first gives

$$- \psi f(- cos\lambda)\ \mathrm{n} + \frac{\psi f}{c}(- cos\lambda)^2\frac{dr}{dt}\mathrm{n} + \frac{\psi f}{c}sin\lambda cos\chi\left(r\frac{d\theta}{dt}\right)(- cos\lambda)\ \mathrm{n}$$

$$- \psi f\ sin\lambda cos\chi\ \S - \frac{\psi f}{c}cos\lambda sin\lambda cos\chi\frac{dr}{dt}\S + \frac{\psi f}{c}sin^2\lambda cos^2\chi\left(r\frac{d\theta}{dt}\right)\S\ .$$

Multiplying through by $(\kappa + 1)cos^{\kappa}\lambda/(2\pi)$ and by the differential of the area, $sin\lambda d\lambda d\chi$, and integrating over λ and χ from 0 to $\pi/2$ and 0 to 2π respectively gives

$$\frac{\psi f}{2\pi}(\kappa + 1)\int_0^{2\pi}\int_0^{\pi/2} cos^{\kappa+1}\lambda sin\lambda d\lambda d\chi\ \mathrm{n}$$

$$+ \frac{\psi f}{2\pi c}(\kappa + 1)\frac{dr}{dt}\int_0^{2\pi}\int_0^{\pi/2} cos^{\kappa+2}\lambda sin\lambda d\lambda d\chi\ \mathrm{n}$$

$$- \frac{\psi f}{2\pi c} (\kappa + 1)\, r \frac{d\theta}{dt} \int_0^{2\pi}\int_0^{\pi/2} cos^{\kappa+1}\lambda sin^2\lambda d\lambda cos\chi d\chi \; \mathbf{n}$$

$$- \frac{\psi f}{2\pi} (\kappa + 1) \int_0^{2\pi}\int_0^{\pi/2} cos^{\kappa}\lambda sin^2\lambda d\lambda cos\chi d\chi \; \S$$

$$- \frac{\psi f}{2\pi c} (\kappa + 1) \frac{dr}{dt} \int_0^{2\pi}\int_0^{\pi/2} cos^{\kappa+1}\lambda sin^2\lambda d\lambda cos\chi d\chi \; \S$$

$$+ \frac{\psi f}{2\pi c} (\kappa + 1)\, r \frac{d\theta}{dt} \int_0^{2\pi}\int_0^{\pi/2} cos^{\kappa}\lambda sin^3\lambda d\lambda cos^2\chi d\chi \; \S \; .$$

Since

$$\int_0^{2\pi} cos\chi d\chi = sin\chi \Big|_0^{2\pi} = 0$$

all the integrals with $cos\chi$ drop out. Evaluating the other integrals of χ leaves

$$\psi f \; (\kappa + 1) \int_0^{\pi/2} cos^{\kappa+1}\lambda sin\lambda d\lambda \; \mathbf{n} + \frac{\psi f}{c} (\kappa + 1) \frac{dr}{dt} \int_0^{\pi/2} cos^{\kappa+2}\lambda sin\lambda d\lambda \; \mathbf{n}$$

$$+ \frac{\psi f}{2\pi c} (\kappa + 1)\, r \frac{d\theta}{dt} \left[\frac{1}{2} sin\chi cos\chi + \frac{1}{2}\chi\right]_0^{2\pi} \int_0^{\pi/2} cos^{\kappa}\lambda sin^3\lambda d\lambda \; \S \; .$$

Further integration and evaluation gives

$$\psi f \; (\kappa + 1) \left[- \frac{cos^{\kappa+2}\lambda}{\kappa + 2}\right]_0^{\pi/2} \mathbf{n} + \frac{\psi f}{c} (\kappa + 1) \frac{dr}{dt} \left[- \frac{cos^{\kappa+3}\lambda}{\kappa + 3}\right]_0^{\pi/2} \mathbf{n}$$

$$+ \frac{\psi f}{2\pi c} (\kappa + 1)\, r \frac{d\theta}{dt} (\pi) \left\{- \frac{sin^2\lambda cos^{\kappa+1}\lambda}{\kappa + 3}\Big|_0^{\pi/2} + \frac{2}{\kappa + 3} \int_0^{\pi/2} cos^{\kappa}\lambda sin\lambda d\lambda \right\} \S \; ;$$

and

$$\psi f \frac{(\kappa + 1)}{(\kappa + 2)} \mathbf{n} + \frac{\psi f}{c} \frac{(\kappa + 1)}{(\kappa + 3)} \frac{dr}{dt} \mathbf{n} + \frac{\psi f}{c} \frac{(\kappa + 1)}{(\kappa + 3)} r \frac{d\theta}{dt} \left[- \frac{\cos^{\kappa+1}\lambda}{\kappa + 1} \right]_0^{\pi/2} \S \ .$$

Finally, the time average values for these terms are

$$\psi f \frac{(\kappa + 1)}{(\kappa + 2)} \mathbf{n} + \frac{\psi f}{c} \frac{(\kappa + 1)}{(\kappa + 3)} \frac{dr}{dt} \mathbf{n} + \frac{\psi f}{c} \frac{1}{\kappa + 3} r \frac{d\theta}{dt} \S \ .$$

Substituting these terms for the last two terms of (49) gives

$$m \frac{d\mathbf{v}}{dt} = f \left(1 - \frac{1}{c} \frac{dr}{dt} \right) \mathbf{n} - (1 - \psi) \frac{f}{c} \frac{dr}{dt} \mathbf{n} - (1 - \psi) \frac{f}{c} r \frac{d\theta}{dt} \S$$

$$+ \psi f \frac{\kappa + 1}{\kappa + 2} \mathbf{n} + \frac{\psi f}{c} \frac{\kappa + 1}{\kappa + 3} \frac{dr}{dt} \mathbf{n} + \frac{\psi f}{c} \frac{1}{\kappa + 3} r \frac{d\theta}{dt} \S \ . \qquad (50)$$

Collecting $\mathbf{n}$ and § terms

$$m \frac{d\mathbf{v}}{dt} = f \left(1 - \frac{1}{c} \frac{dr}{dt} - \frac{(1 - \psi)}{c} \frac{dr}{dt} + \psi \frac{\kappa + 1}{\kappa + 2} + \frac{\psi}{c} \frac{\kappa + 1}{\kappa + 3} \frac{dr}{dt} \right) \mathbf{n}$$

$$- \frac{fr}{c} \frac{d\theta}{dt} \left((1 - \psi) - \frac{\psi}{\kappa + 3} \right) \S \ .$$

Regrouping terms in the brackets and substituting for f

$$m \frac{d\mathbf{v}}{dt} = \frac{m\alpha}{r^2} \left[c \left(1 + \psi \frac{\kappa + 1}{\kappa + 2} \right) - \frac{dr}{dt} \left(2 - \psi - \psi \frac{\kappa + 1}{\kappa + 3} \right) \right] \mathbf{n}$$

$$- \frac{m\alpha}{r^2} r \frac{d\theta}{dt} \left(1 - \frac{\psi\kappa + \psi 3 + \psi}{\kappa + 3} \right) \S \ ,$$

and after some more algebraic manipulations

$$m \frac{d\mathbf{v}}{dt} = \frac{m\alpha}{r^2} \left[c \left(1 + \psi \frac{\kappa + 1}{\kappa + 2} \right) - 2 \frac{dr}{dt} \left(1 - \psi \frac{\kappa + 2}{\kappa + 3} \right) \right] \mathbf{n}$$

$$- \frac{m\alpha}{r^2} r \frac{d\theta}{dt} \left(1 - \psi \frac{\kappa + 4}{\kappa + 3} \right) \S \ . \qquad (51)$$

Defining the following constants will make (51) more compact. Let

$$\alpha_0 \equiv \alpha\left(1 + \psi \frac{\kappa + 1}{\kappa + 2}\right) , \tag{52}$$

$$\alpha_1 \equiv \alpha\left(1 - \psi \frac{\kappa + 2}{\kappa + 3}\right) , \tag{53}$$

and

$$\beta \equiv \alpha\left(1 - \psi \frac{\kappa + 4}{\kappa + 3}\right) . \tag{54}$$

Then

$$m \frac{d\mathbf{v}}{dt} = \frac{m\alpha_0 c}{r^2} \mathrm{n} - \frac{2m\alpha_1}{r^2} \frac{dr}{dt} \mathrm{n} - \frac{m\beta}{r^2} r \frac{d\theta}{dt} \S . \tag{55}$$

Using the expression for radial acceleration and including the solar gravitational attraction,

$$m\left[\frac{d^2r}{dt^2} - r\left(\frac{d\theta}{dt}\right)^2\right] = \frac{m\alpha_0 c}{r^2} - \frac{2m\alpha_1}{r^2} \frac{dr}{dt} - \frac{GMm}{r^2} .$$

Let $\mu_0 \equiv - GM$. Then

$$\frac{d^2r}{dt^2} - r\left(\frac{d\theta}{dt}\right)^2 = - \frac{\mu_0 - \alpha_0 c}{r^2} - \frac{2\alpha_1}{r^2} \frac{dr}{dt} . \tag{56}$$

Again, applying the expression for the angular acceleration

$$m \frac{1}{r} \frac{d}{dt}\left(r^2 \frac{d\theta}{dt}\right) = - \frac{m\beta}{r^2} r \frac{d\theta}{dt}$$

and

$$\frac{1}{r} \frac{d}{dt}\left(r^2 \frac{d\theta}{dt}\right) = - \frac{\beta}{r^2} r \frac{d\theta}{dt} . \tag{57}$$

Equations (56) and (57) are Lyttleton's equations of motion for the case of partially reflected radiation. If $\psi = 0$, the case where all the energy is absorbed, these equations revert to the same expressions as those obtained by Robertson. On the other hand, if $\psi = 1$ and $\kappa = \infty$, all the incident radiation is reflected directly back toward the sun and the drag force is eliminated. The radial component of the acceleration becomes

$$- \frac{\mu_0 - 2\alpha c}{r^2}$$

showing that the radiation pressure has been doubled.

The general results of including the effects of the reflected light are to increase the repulsion effect of the radiation, to reduce the magnitude of the Doppler effect, and to reduce the amount of tangential drag on the particle.

PERTURBATIONS OF THE ORBIT

In order to facilitate the making of approximations in the following calculations it is helpful at this time to obtain an order of magnitude estimate of the various terms in (56) and (57). For this purpose they can be rewritten as

$$\frac{d^2r}{dt^2} - r\left(\frac{d\theta}{dt}\right)^2 = - \frac{GM}{r^2} + \frac{\alpha c}{r^2} - \frac{2\alpha}{r^2}\frac{dr}{dt}$$

and

$$\frac{1}{r}\frac{d}{dt}\left(r^2 \frac{d\theta}{dt}\right) = - \frac{\beta r}{r^2}\frac{d\theta}{dt} .$$

The dimensions of α and β are in cm^2/s in c.g.s. units as stated in (32). GM has units cm^3/s^2 and αc also is cm^3/s^2. The value of α/r^2 at 1 AU is given in (32). The magnitudes of the velocity components,

$(dr)/(dt)$ and $r(d\theta)/(dt)$ are approximately 3×10^6 *cm/s* if the particle is in an orbit of high eccentricity near the earth's orbit.[43] The values of these terms are then

$$GM = \left(6.68\times10^{-8}\ dyne\text{-}cm^2\ g^{-2}\right)\left(1.991\times10^{33}\ g\right)$$

$$= 1.3\times10^{26}\ cm^3/s^2\ ,$$

$$\alpha c = \left(\frac{2.54\times10^{11}}{a\rho}\ \frac{cm^2}{s}\right)\left(2.998\times10^{10}\ cm/s\right)$$

$$= \frac{7.6\times10^{21}}{a\rho}\ cm^3/s^2\ ,$$

$$2\alpha\ \frac{dr}{dt} = 2\left(\frac{2.54\times10^{11}}{a\rho}\ \frac{cm^2}{s}\right)\left(3\times10^{6}\ cm/s\right)$$

$$= \frac{1.5\times10^{18}}{a\rho}\ cm^3/s^2\ ,$$

$$\beta r\ \frac{d\theta}{dt} = \left(\frac{2.54\times10^{11}}{a\rho}\ \frac{cm^2}{s}\right)\left(3\times10^{6}\ cm/s\right)$$

$$= \frac{7.5\times10^{17}}{a\rho}\ cm^3/s^2\ .$$

From these rough values it can be seen that the radiation pressure is about 5×10^3 times as great as the inward force, $2\alpha(dr/dt)/r^2$, due to the Doppler effect. It is also close to 10^4 times as great as the tangential drag acting on the particle which depends directly on the velocity of the particle. As will be shown in the following derivations, the velocity varies as $r^{-1/2}$, and even when the particle is quite close to the sun the tangential drag is still approximately 10^3 times less than the outward

---

[43] R. A. Lyttleton, p. 124.

pressure of the radiation.

The following theory will develop the equations which describe the motion of a particle in the attracting field of a center of force. Let this force depend in some way on the distance r from the center of the field. As a reference to the earlier portion of this work consider the particle to be at rest in Σ_E at the instant E. Thus, there is no tangential drag acting at this time. The following equations have already been seen to hold true

$$m\left[\frac{d^2r}{dt^2} - r\left(\frac{d\theta}{dt}\right)^2\right] = F(r) \tag{58}$$

and

$$mr^2 \frac{d\theta}{dt} = mh \cdot \tag{59}$$

where h is defined as the angular momentum per unit mass.

Let $\sigma \equiv 1/r$. Then

$$mr^2 \frac{d\theta}{dt} = \frac{m}{\sigma^2}\frac{d\theta}{dt} = mh$$

and

$$\frac{d\theta}{dt} = h\sigma^2 \ . \tag{60}$$

Also,

$$\frac{dr}{dt} = \frac{d(1/\sigma)}{dt} = -\frac{1}{\sigma^2}\frac{d\sigma}{dt} = -\frac{1}{\sigma^2}\frac{d\sigma}{d\theta}\frac{d\theta}{dt} \ ;$$

$$\frac{dr}{dt} = -\frac{1}{\sigma^2}\frac{d\sigma}{d\theta}h\sigma^2 = -h\frac{d\sigma}{d\theta} \ . \tag{61}$$

In the same way

$$\frac{d^2r}{dt^2} = \frac{d}{dt}\left(- h \frac{d\sigma}{d\theta}\right) = - \frac{dh}{dt}\frac{d\sigma}{dt} - h \frac{d^2\sigma}{dt^2}$$

and

$$\frac{d^2r}{dt^2} = - h \frac{d^2\sigma}{dt^2}$$

since the angular momentum is constant at E. Then

$$\frac{d^2r}{dt^2} = - h \frac{d^2\sigma}{d\theta^2}\frac{d\theta}{dt} = - h^2\sigma^2 \frac{d^2\sigma}{d\theta^2} . \tag{62}$$

Substitution of these values in (58) gives

$$m\left(- h^2\sigma^2 \frac{d^2\sigma}{d\theta^2} - \frac{1}{\sigma} h^2\sigma^4\right) = F(1/\sigma)$$

or

$$\frac{d^2\sigma}{d\theta^2} + \sigma = - \frac{F(1/\sigma)}{mh^2\sigma^2} . \tag{63}$$

If it is assumed that the force varies as an integral power, n, of the distance, then

$$F(r) = ar^n \quad \text{and} \quad F(1/\sigma) = a\sigma^{-n}$$

where a is a proportionality constant. Equation (63) becomes

$$\frac{d^2\sigma}{d\theta^2} + \sigma = - \frac{a\sigma^{-n-2}}{mh^2} .$$

In order to obtain an integrable form of this equation, multiply both sides by $2(d\sigma/d\theta)$;

$$2\left(\frac{d\sigma}{d\theta}\right)\left(\frac{d^2\sigma}{d\theta^2} + \sigma\right) = - \frac{2a\sigma^{-n-2}}{mh^2}\frac{d\sigma}{d\theta}$$

$$2\frac{d\sigma}{d\theta}\frac{d^2\sigma}{d\theta^2} + 2\sigma\frac{d\sigma}{d\theta} = -b\sigma^{-n-2}\frac{d\sigma}{d\theta}$$

where $b \equiv 2a/(mh^2)$. The term on the left can be rewritten as

$$\frac{d}{d\theta}\left[\left(\frac{d\sigma}{d\theta}\right)^2 + \sigma^2\right] .$$

Integrating gives

$$\int d\left[\left(\frac{d\sigma}{d\theta}\right)^2 + \sigma^2\right] = -b\int \sigma^{-n-2}\, d\sigma$$

and

$$\left(\frac{d\sigma}{d\theta}\right)^2 + \sigma^2 = \frac{b\sigma^{-n-1}}{n+1} + C \qquad \text{where } n \neq -1 . \tag{64}$$

Going back to (63), if $F(r) = -GMm/r^2$, then $F(1/\sigma) = -GMm\sigma^2$ and

$$\frac{d^2\sigma}{d\theta^2} + \sigma = \frac{GM}{h^2} .$$

The general solution for this expression is of the form

$$\sigma = A\, cos\theta .$$

For this particular case

$$\sigma = A\, cos(\theta - \theta_0) + GM/h^2 \tag{65}$$

where θ_0 is the phase angle. Since $r = 1/\sigma$,

$$r = \frac{h^2/GM}{1 + (Ah^2/GM)cos(\theta - \theta_0)} . \tag{66}$$

The standard equation for a conic section is

$$r = \frac{p}{1 + e\, cos(\theta - \theta_0)} \tag{67}$$

where e is the eccentricity and p/e is the distance from the focus to the

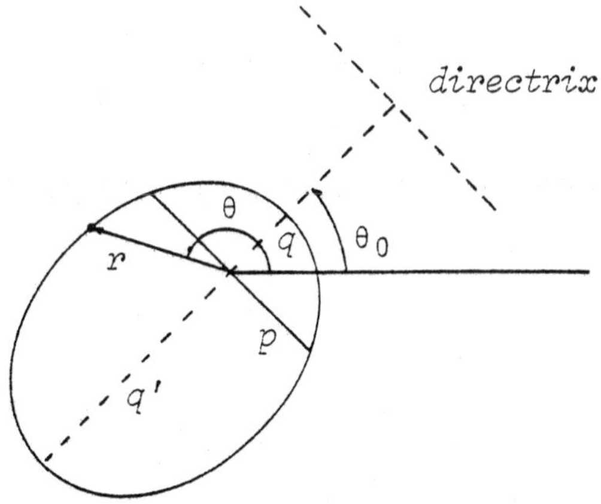

Figure (7). Elements of elliptical orbit.

directrix of the conic section. Comparing (66) and (67) shows that

$$p = \frac{h^2}{GM} \quad \text{and} \quad e = \frac{Ah^2}{GM} . \tag{68}$$

If the inverse square law is applied to (64), then

$$\left(\frac{d\sigma}{d\theta}\right)^2 + \sigma^2 = - \frac{2a}{mh^2}\sigma + C = \frac{2GM}{h^2}\sigma + C \tag{69}$$

since $n = -2$, and $a = -GMm$. The radial and angular components of the velocity can be written as

$$\frac{dr}{dt} = - h \frac{d\sigma}{d\theta} \quad \text{and} \quad r \frac{d\theta}{dt} = \frac{1}{\sigma}(h\sigma^2) = h\sigma .$$

The speed of the object is

$$\left[\left(\frac{dr}{dt}\right)^2 + \left(r \frac{d\theta}{dt}\right)^2\right]^{1/2} = \left[\left(h \frac{d\sigma}{d\theta}\right)^2 + h^2\sigma^2\right]^{1/2} . \tag{70}$$

Also, since

$$\frac{dV}{dr} = - F(r) ,$$

and

$$\int dV = GMm \int \frac{dr}{r^2} ,$$

the value of the potential energy can be expressed as

$$V = - \frac{GMm}{r} + V_0$$

or

$$V = - \frac{GMm}{r} \qquad \text{if } V_0 \equiv 0 \ . \tag{71}$$

From (70) and (71) it can be seen that (69) is an expression of the total energy per unit mass since

$$\frac{m}{2} \left[h^2 \left(\frac{d\sigma}{d\theta} \right)^2 + h^2\sigma^2 \right] - GMm\sigma = \frac{mCh^2}{2} \tag{72}$$

or

$$K.E. + P.E. = \mathit{Total\ Energy} ,$$

and

$$C = \frac{2E}{mh^2}$$

where E is the total energy. At the ends of the transverse axis of the ellipse, $dr/dt = - h(d\sigma/d\theta) = 0$. Applying this to (72) gives

$$\sigma^2 - \frac{2GMm}{h^2} - \frac{2E}{mh^2} = 0 \ .$$

The value of σ can be determined by the quadratic formula;

$$\sigma = \frac{\dfrac{2GMm}{h^2} \pm \left[\left(\dfrac{2GM}{h^2} \right)^2 + 4 \left(\dfrac{2E}{mh^2} \right) \right]^{1/2}}{2} ,$$

$$\sigma = \frac{GM}{h^2}\left[1 \pm \left(1 + \frac{2h^2 E}{mG^2M^2}\right)^{1/2}\right] .$$

The value of σ which corresponds to the end of the transverse axis closest to the center of force (the focus) is

$$\sigma = \frac{GM}{h^2}\left[1 + \left(1 + \frac{2h^2 E}{mG^2M^2}\right)^{1/2}\right] , \tag{73}$$

and for the far end

$$\sigma = \frac{GM}{h^2}\left[1 - \left(1 + \frac{2h^2 E}{mG^2M^2}\right)^{1/2}\right] . \tag{74}$$

Comparing this form with (65), it is easily seen that for the case of the near end $\theta - \theta_0 = 0$, and for the far end $\theta - \theta_0 = \pi$. Thus, for the near end

$$\sigma = \frac{GM}{h^2} + A , \quad \text{and for the far end} \quad \sigma = \frac{GM}{h^2} - A .$$

Comparision of these expressions with (73) and (74) shows that

$$A = \frac{GM}{h^2}\left(1 + \frac{2h^2 E}{mG^2M^2}\right)^{1/2} .$$

Substituting this value in (68) gives

$$e = \left(1 + \frac{2h^2 E}{mG^2M^2}\right)^{1/2} . \tag{75}$$

Since this study is concerned with particles in closed orbits around the sun, their orbits will be elliptic except for the special case of circular orbits. Thus, $e < 1$ and $E < 0$.

If the two special cases for the length of the radius vector considered above are labeled q and q', where q is the distance to the near end and q' is the distance to the far end, the following two expressions result from (67):

$$q = \frac{p}{1 + e} \quad \text{and} \quad q' = \frac{p}{1 - e} \; . \tag{76}$$

If the total length of the transverse axis (major axis) is defined as $2a$, then

$$q + q' = 2a \tag{77}$$

and a is the length of the semimajor axis. Combining (76) and (77) gives

$$p = a(1 - e^2) \; . \tag{78}$$

Using (78), (67) can be rewritten as

$$r = \frac{a(1 - e^2)}{1 + e \cos(\theta - \theta_0)} \; . \tag{79}$$

Equation (68) stated that $p = h^2/GM$, thus by substituting (78) for p

$$h = \left(GMa(1 - e^2)\right)^{1/2} \; . \tag{80}$$

Substituting this in (75) gives

$$e = \left(1 + \frac{2E\left(GMa(1 - e^2)\right)}{mG^2M^2}\right)^{1/2} ,$$

$$e^2 = 1 + \frac{2a(1 - e^2)E}{mGM} ,$$

and

$$E = \frac{(e^2 - 1)GMm}{2a(1 - e^2)} = -\frac{GMm}{2a} \; . \tag{81}$$

With this expression, the square of the orbital speed at some distance r from the center of force is easily obtained using (72), where it was seen that

$$E = \frac{mv^2}{2} - \frac{GMm}{r} = - \frac{GMm}{2a} .$$

Then

$$v^2 = GM \left(\frac{2}{r} - \frac{1}{a} \right)$$

and the kinetic energy per unit mass is

$$\frac{v^2}{2} = GM \left(\frac{1}{r} - \frac{1}{2a} \right) .$$

This can also be written as

$$\frac{1}{2} \left[\left(\frac{dr}{dt} \right)^2 + r^2 \left(\frac{d\theta}{dt} \right)^2 \right] = \mu \left(\frac{1}{r} - \frac{1}{2a} \right) \qquad (82)$$

since in this problem the radial force will be the net effect of solar gravitational attraction and radiation pressure. Differentiating (82) with respect to time

$$\frac{dr}{dt}\frac{d^2r}{dt^2} + r\frac{dr}{dt}\left(\frac{d\theta}{dt}\right)^2 + r^2\frac{d\theta}{dt}\frac{d^2\theta}{dt^2} = -\frac{\mu}{r^2}\frac{dr}{dt} + \frac{\mu}{2a^2}\frac{da}{dt}$$

or

$$\frac{dr}{dt}\left[\frac{d^2r}{dt^2} - r\left(\frac{d\theta}{dt}\right)^2\right] + 2r\frac{dr}{dt}\left(\frac{d\theta}{dt}\right)^2 + r^2\frac{d\theta}{dt}\frac{d^2\theta}{dt^2} = -\frac{\mu}{r^2}\frac{dr}{dt} + \frac{\mu}{2a^2}\frac{da}{dt} .$$

By (56) and (57),

$$\frac{dr}{dt}\left(-\frac{\mu}{r^2} - \frac{2\alpha_1}{r^2}\frac{dr}{dt}\right) + r\frac{d\theta}{dt}\left(-\frac{\beta}{r^2}r\frac{d\theta}{dt}\right) = -\frac{\mu}{r^2}\frac{dr}{dt} + \frac{\mu}{2a^2}\frac{da}{dt} .$$

Rearranging leaves

$$\frac{\mu}{2a^2}\frac{da}{dt} = -\frac{2\alpha_1}{r^2}\left(\frac{dr}{dt}\right)^2 - \beta\left(\frac{d\theta}{dt}\right)^2 . \qquad (83)$$

Notice that the drag force is now considered to be operating since the expression for the angular acceleration was not set equal to zero. Thus, the particle is now moving with respect to the origin of Σ_E.

The purpose behind these manipulations is to obtain expressions of the changes in the orbital elements with time, especially the semimajor axis and the eccentricity. With some substitution in (83) the equation for the change in the semimajor axis can be obtained. The values α_1 and β have already been shown to be very small when compared to the other forces acting on the particle. For this reason it is acceptable to consider the elements of the orbit to be constant for a relatively short time such as one period of the orbit. This will allow a simpler evaluation of (83) which will still be a good approximation in terms of the inaccuracy due to knowing little about the particle itself.

It has already been shown that

$$r = \frac{a(1 - e^2)}{1 + e\cos\theta} .$$

Differentiating with respect to time gives

$$\frac{dr}{dt} = -\frac{a(1 - e^2)(-e\sin\theta\, d\theta/dt)}{(1 + e\cos\theta)^2}$$

$$= \frac{re\sin\theta\,(d\theta/dt)}{(1 + e\cos\theta)} .$$

Since $d\theta/dt = h/r^2$,

$$\frac{dr}{dt} = \frac{reh\sin\theta}{r^2(1 + e\cos\theta)} = \frac{eh\sin\theta}{a(1 - e^2)} . \qquad (84)$$

The area swept out by a radius vector during some time, t, is

$$A = 1/2\, ht .$$

Substituting (80) for h gives

$$A = 1/2\left(\mu a(1 - e^2)\right)^{1/2} t$$

where $\mu = GM - \alpha c$. The area of an ellipse is πab where b is the semi-minor axis. The radius vector will sweep out the total area of the ellipse during one period. Thus

$$\pi ab = \pi a^2\left(1 - e^2\right)^{1/2} = 1/2\left(\mu a(1 - e^2)\right)^{1/2} P$$

and

$$P = \frac{2\pi a^{3/2}}{\sqrt{\mu}} \tag{85}$$

where $b = a\left(1 - e^2\right)^{1/2}$. Since $P = 2\pi(d\theta/dt)^{-1}$,

$$\frac{2\pi}{d\theta/dt} = \frac{2\pi a^{3/2}}{\sqrt{\mu}}$$

and

$$\mu = a^3\left(\frac{d\theta}{dt}\right)^2 . \tag{86}$$

Also, since $r^2 = ab$,

$$h = ab\left(\frac{d\theta}{dt}\right) . \tag{87}$$

Now, going back to (84), the value for h can be substituted by (87) giving

$$\frac{dr}{dt} = \frac{eab\ sin\theta}{a(1 - e^2)}\left(\frac{d\theta}{dt}\right) .$$

Squaring this

$$\left(\frac{dr}{dt}\right)^2 = \frac{e^2b^2\ sin^2\theta}{(1 - e^2)^2}\left(\frac{d\theta}{dt}\right)^2 ,$$

and substituting values for b and $d\theta/dt$

$$\left(\frac{dr}{dt}\right)^2 = \frac{e^2a^2(1 - e^2)sin^2\theta}{(1 - e^2)^2}\left(\frac{\mu}{a^3}\right) ,$$

which simplifies to

$$\left(\frac{dr}{dt}\right)^2 = \frac{e^2\mu sin^2\theta}{a(1 - e^2)} . \quad (88)$$

Expression (83) has the term $(dr/dt)^2/r^2$ which is equivalent to

$$\frac{e^2\mu\ sin^2\theta}{a(1 - e^2)}\left[\frac{(1 + e\ cos\theta)^2}{a^2(1 - e^2)^2}\right] = \frac{e^2\mu\ sin^2\theta}{a^3(1 - e\ )^3}(1 + e\ cos\theta)^2 .$$

Again, since $r^2 = ab$,

$$\frac{a^2(1 - e^2)^2}{(1 + e\ cos\theta)^2} = a^2(1 - e^2)^{1/2} ,$$

and

$$(1 - e^2)^{3/2} = (1 + e\ cos\theta)^2 . \quad (89)$$

Thus

$$\frac{(dr/dt)^2}{r^2} = \frac{e^2\mu\ sin^2\theta}{a^3(1 - e^2)^{3/2}} . \quad (90)$$

Equation (83) can now be rewritten as

$$\frac{\mu}{2a^2}\left(\frac{da}{dt}\right) = -\frac{2\alpha_1e^2\mu\ sin^2\theta}{a^3(1 - e^2)^{3/2}} - \frac{\mu\beta}{a^3} .$$

Multiplying through by a^3/μ leaves

$$\frac{a}{2}\left(\frac{da}{dt}\right) = -\frac{2\alpha_1e^2sin^2\theta}{(1 - e^2)^{3/2}} - \beta$$

or

$$\frac{a}{2}\left(\frac{da}{dt}\right) = -\frac{2\alpha_1e^2sin^2\theta}{(1 - e^2)^{3/2}} - \frac{\beta(1 + e\ cos\theta)^2}{(1 - e^2)^{3/2}} .$$

Integrating both sides over θ from 0 to 2π

$$\frac{a}{2}\left(\frac{da}{dt}\right)\int_0^{2\pi} d\theta = -\frac{2\alpha_1 e^2}{(1-e^2)^{3/2}}\int_0^{2\pi} \sin^2\theta d\theta - \frac{\beta}{(1-e^2)^{3/2}}\int_0^{2\pi} (1 + e\cos\theta)^2 d\theta$$

and evaluating the integrals gives

$$(2\pi)\frac{a}{2}\left(\frac{da}{dt}\right) = -\frac{2\alpha_1 e^2}{(1-e^2)^{3/2}}(\pi) - \frac{\beta}{(1-e^2)^{3/2}}(2\pi + \pi e^2)$$

$$= -2\pi\left(\frac{\alpha_1 e^2}{(1-e^2)^{3/2}} - \frac{\beta}{(1-e^2)^{3/2}}(1 + 1/2e^2)\right) .$$

Dividing through by 2π leaves

$$\frac{a}{2}\left(\frac{da}{dt}\right) = -\frac{\alpha_1 e^2 + \beta(1 + 1/2e^2)}{(1-e^2)^{3/2}} . \qquad (91)$$

If there is no reflection, α_1 and β are equal to α and (91) becomes

$$\frac{da}{dt} = -\frac{\alpha(2 + 3e^2)}{a(1-e^2)^{3/2}} \qquad (92)$$

which is the expression reported by Robertson and utilized by Wyatt and Whipple in their paper.

It was shown earlier that

$$h = \left(\mu a(1-e^2)\right)^{1/2} .$$

Squaring both sides

$$h^2 = \mu a(1-e^2) ,$$

and taking the derivative with respect to t

$$2h\frac{dh}{dt} = \mu\frac{da}{dt}(1-e^2) + \mu(-2e)\frac{de}{dt}a$$

or by rearranging,

$$(\mu a)\,2e\frac{de}{dt} = \mu\frac{da}{dt}(1-e^2) - 2h\frac{dh}{dt} .$$

Substituting for da/dt, h and dh/dt gives

$$(\mu a)\ 2e \frac{de}{dt} = -\frac{\mu(1-e^2)}{(1-e^2)^{3/2}}\left(\frac{2}{a}\right)\left(\alpha_1 e^2 + \beta(1 + 1/2e^2)\right) + 2\frac{d\theta}{dt} ab\beta \frac{d\theta}{dt} .$$

Simplifying and combining terms

$$(\mu a)\ 2e \frac{de}{dt} = -\frac{\mu(2\alpha_1 e^2 + 2\beta + \beta e^2)}{a(1-e^2)^{1/2}} + \frac{2\mu a^2(1-e^2)^{1/2}\beta}{a^3} ,$$

$$2e\frac{de}{dt} = -\frac{2\alpha_1 e^2 + 2\beta + \beta e^2}{a^2(1-e^2)^{1/2}} + \frac{2\beta(1-e^2)}{a^2(1-e^2)^{1/2}}$$

$$= -\frac{2\alpha_1 e^2 + 2\beta + \beta e^2 - 2\beta + 2\beta e^2}{a(1-e^2)^{1/2}}$$

$$= -\frac{2\alpha_1 e^2 + 3\beta e^2}{a^2(1-e^2)^{1/2}} ,$$

and

$$\frac{2}{e}\frac{de}{dt} = -\frac{2\alpha_1 + 3\beta}{a^2(1-e^2)^{1/2}} . \qquad (93)$$

Again, if no reflection occurs this can be rewritten as

$$\frac{de}{dt} = -\frac{5\alpha e}{2a^2(1-e^2)^{1/2}} \qquad (94)$$

which matches Robertson's result.

TIME OF INFALL

Two basic studies have been made to determine the lifetime of a dust particle given the initial values of the eccentricity and semimajor axis of its orbit. The first was done by Wyatt and Whipple in 1950 using the results of Robertson's investigation. Lyttleton used his own modified equations to study this aspect of the problem in 1976. A demonstration of the validity of their results will be performed in this section.

The calculations for the case of total absorption proceed in the following manner. Divide (92) by (94),

$$\frac{da/dt}{de/dt} = - \frac{\alpha(2 + 3e^2)}{a(1 - e^2)^{3/2}} \left[- \frac{2a^2(1 - e^2)^{1/2}}{5\alpha e} \right]$$

which gives

$$\frac{da}{de} = \frac{2a(2 + 3e^2)}{5e(1 - e^2)} . \tag{95}$$

Separating variables,

$$\frac{da}{a} = \frac{2}{5} \left[\frac{2}{e(1 - e^2)} de + \frac{3e}{1 - e^2} de \right] .$$

Let $e \equiv sin\theta$ then $de = cos\theta d\theta$ and,

$$\frac{da}{a} = \frac{4}{5} \frac{cos\theta d\theta}{sin\theta cos^2\theta} + \frac{6}{5} \frac{sin\theta cos\theta}{cos^2\theta} d\theta .$$

Integrating both sides

$$\int \frac{da}{a} = \frac{4}{5} \int \frac{d\theta}{sin\theta cos\theta} + \frac{6}{5} \int tan\theta d\theta ,$$

which is

$$ln\ a = 4/5\ ln\ tan\theta - 6/5\ ln\ cos\theta + ln\ C$$

or

$$a = \frac{C\ tan^{4/5}\theta}{cos^{6/5}\theta} .$$

Substituting the original terms gives

$$a = \frac{C\left(e/(1 - e^2)^{1/2}\right)^{4/5}}{\left((1 - e^2)^{1/2}\right)^{6/5}} = \frac{C\ e^{4/5}}{(1 - e^2)^{3/5}(1 - e^2)^{2/5}} ,$$

$$a = \frac{C\, e^{4/5}}{1 - e^2} . \qquad (96)$$

If the original values of a and e are a_0 and e_0 respectively then C can be evaluated

$$C = a_0 e_0^{-4/5}\left(1 - e_0^2\right) .$$

Equation (96) can now be used to write (94) as a function of e and t only.

$$\frac{de}{dt} = -\frac{5\alpha e}{2\dfrac{C^2 e^{8/5}}{(1 - e^2)^2}(1 - e^2)^{1/2}} = -\frac{5\alpha(1 - e^2)^{3/2}}{2C^2 e^{3/5}} .$$

Separating variables gives

$$dt = -\frac{2C^2}{5\alpha}\frac{e^{3/5}}{(1 - e^2)^{3/2}}\, de ,$$

and by integrating

$$t - t_0 = -\frac{2C^2}{5\alpha}\int_{e_0}^{e}\frac{e^{3/5}}{(1 - e^2)^{3/2}}\, de . \qquad (97)$$

The value of α was given in (32). Converting this to $AU^2/year$

$$\alpha = \frac{1}{a\rho}(2.54\times10^{11})\frac{cm^2}{s}\left[\frac{3.156\times10^7\ s/year}{(1.496\times10^{13})^2 cm^2/AU^2}\right] ,$$

$$\alpha = \frac{3.58\times10^{-8}}{a\rho}\frac{AU^2}{year}$$

where a is now being used as the radius of the particle. Then (97) becomes

$$(t - t_0)_{years} = 1.12\times10^7\ a\rho C^2\int_{e}^{e_0} e^{3/5}(1 - e^2)^{-3/2} de \qquad (98)$$

where $e_0 > e$ and C is expressed in AU. To obtain the time of infall let the lower limit equal zero, and evaluate the integral.

The procedure to find the lifetime of the particle with (91) and (93) is similar to the one above for the case of total absorption. Let $x = a^2$ and $y = e^2$. Then,

$$\frac{dx}{dt} = 2a\frac{da}{dt} \quad \text{and} \quad \frac{dy}{dt} = 2e\frac{de}{dt} .$$

Equation (91) can be rewritten

$$4\left(\frac{a}{2}\frac{da}{dt}\right) = -\frac{4\alpha_1 y + 4\beta(1 + 1/2y)}{(1 - y)^{3/2}} ,$$

and

$$\frac{dx}{dt} = -\frac{(4\alpha_1 + 2\beta)y + 4\beta}{(1 - y)^{3/2}} . \tag{99}$$

Equation (93) can be rewritten

$$e^2\left(\frac{2}{e}\frac{de}{dt}\right) = -\frac{(2\alpha_1 + 3\beta)e^2}{a^2(1 - e^2)^{1/2}} ,$$

or

$$x\frac{dy}{dt} = -\frac{(2\alpha_1 + 3\beta)y}{(1 - y)^{1/2}} . \tag{100}$$

Dividing (99) by (100),

$$\frac{1}{x}\frac{dx/dt}{dy/dt} = \frac{(4\alpha_1 + 2\beta)y + 4\beta}{(1 - y)^{3/2}} \frac{(1 - y)^{1/2}}{(2\alpha_1 + 3\beta)y} ,$$

$$\frac{1}{x}\frac{dx}{dy} = \frac{(4\alpha_1 + 2\beta)y + 4\beta}{(1 - y)(2\alpha_1 + 3\beta)y} = \frac{(4\alpha_1 + 6\beta)y + 4\beta(1 - y)}{(1 - y)(2\alpha_1 + 3\beta)y} ,$$

$$\frac{1}{x}\frac{dx}{dy} = \frac{2}{1 - y} + \frac{\gamma}{y} \quad \text{where } \gamma \equiv \frac{4\beta}{2\alpha_1 + 3\beta} . \tag{101}$$

By separating variables and integrating

$$\int_{x_0}^{x} \frac{dx}{x} = -\int_{y_0}^{y} \frac{2}{1 - y}(-dy) + \int_{y_0}^{y} \frac{\gamma}{y} dy .$$

Evaluating the integrals

$$ln\ x - ln\ x_0 = 2ln\ (1 - y_0) - 2ln\ (1 - y) + \gamma ln\ y - \gamma ln\ y_0 ,$$

$$ln\ \frac{x}{x_0} = ln\ \frac{(1 - y_0)^2}{(1 - y)^2} \frac{y^\gamma}{y_0^\gamma} ,$$

or

$$\frac{x}{x_0} = \frac{(1 - y_0)^2}{(1 - y)^2} \frac{y^\gamma}{y_0^\gamma} ,$$

and

$$x = \frac{x_0(1 - y_0)^2}{y_0^\gamma} \frac{y^\gamma}{(1 - y)^2} .$$

Let

$$K \equiv \frac{x_0(1 - y_0)^2}{y_0^\gamma} = \frac{a_0(1 - e_0^2)^2}{e_0^{2\gamma}} .$$

Then,

$$x = \frac{K\ y^\gamma}{(1 - y)^2} . \qquad (102)$$

Substituting (102) into (100) in order to eliminate x leaves

$$\frac{K\ y^\gamma}{(1 - y)^2} \frac{dy}{dt} = -\frac{(2\alpha_1 + 3\beta)y}{(1 - y)^{1/2}} .$$

Separation of variables gives

$$-\frac{2\alpha_1 + 3\beta}{K} dt = \frac{y^{\gamma - 1}}{(1 - y)^{3/2}} dy ,$$

and by integrating

$$-\frac{2\alpha_1 + 3\beta}{K}\int_{t_o}^{t} dt = \int_{y_o}^{y} y^{\gamma-1}(1 - y)^{-3/2}dy ,$$

the result is

$$t - t_o = \frac{K}{2\alpha_1 + 3\beta}\int_{y}^{y_o} y^{\gamma-1}(1 - y)^{-3/2}dy . \tag{103}$$

Note that if $\gamma = 4/5$, K is equal to the constant C^2 used for the case of total absorption. K can also be expressed as

$$K = \frac{C^2}{e_o^{2\gamma-8/5}} . \tag{104}$$

The expression, $2\alpha_1 + 3\beta$, can be rewritten using the definitions of α_1 and β. Thus

$$2\alpha_1 + 3\beta = 2\alpha\left(1 - \psi\frac{\kappa + 2}{\kappa + 3}\right) + 3\alpha\left(1 - \psi\frac{\kappa + 4}{\kappa + 3}\right)$$

$$= \alpha\left(5 - \psi\frac{2\kappa + 4 + 3\kappa + 12}{\kappa + 3}\right)$$

$$2\alpha_1 + 3\beta = 5\alpha\left(1 - \psi\frac{\kappa + 16/5}{\kappa + 3}\right) . \tag{105}$$

Substituting (104) and (105) in (103) gives

$$t - t_o = \frac{C^2}{5\alpha e_o^{2\gamma-8/5}\left(1 - \psi\frac{\kappa + 16/5}{\kappa + 3}\right)}\int_{y}^{y_o} y^{\gamma-1}(1 - y)^{-3/2}dy ,$$

or

$$t - t_o = \frac{5.57\times10^6\ a\rho C^2}{e_o^{2\gamma-8/5}\left(1 - \psi\frac{\kappa + 16/5}{\kappa + 3}\right)}\int_{y}^{y_o} y^{\gamma-1}(1 - y)^{-3/2}dy . \tag{106}$$

To obtain the time of infall, let y = 0 for the lower limit of the integral in (106), and evaluate it. For this case of partial reflection, ψ, γ and κ must also be defined in order to calculate a specific particle lifetime.

PART IV

EXTINCTION OF INTERPLANETARY DUST

The calculation for time of infall using (98) is straightforward. A numerical method was used to evaluate the integral for interesting values of the original eccentricity and a lower limit of 0.[44] The values of this integral are listed in Table IV where it is represented as $G(e_0)$.

A bit more work must be done to obtain a value using (106). The symbol ψ represents the fraction of the incident radiation which is reflected, *i.e.* the albedo. An average particle in the solar system has a surface albedo of 0.1 according to Table I. This is the value which will be used throughout the remainder of this thesis.

The values of κ and γ must also be defined. The relationship between them is given in (101). The denominator is expressed in (105). Thus,

$$\gamma = \frac{4\alpha\left(1 - \psi\,\dfrac{\kappa + 4}{\kappa + 3}\right)}{5\alpha\left(1 - \psi\,\dfrac{\kappa + 16/5}{\kappa + 3}\right)} = \frac{4}{5}\,\frac{\kappa + 3 - \psi(\kappa + 4)}{\kappa + 3 - \psi(\kappa + 16/5)}\;. \qquad (107)$$

It is immediately apparent that for normal ranges of ψ, γ will have values less than 4/5. The value of γ is 4/5 when $\psi = 0$. This is the case of total absorption, and (106) gives particle lifetimes which are identical to those obtained using (98). In order to make these results comparable

---

[44]This was done using the Hewlett Packard 9825A calculator. The software component labeled "General Utility Routines" contained the program entitled *Numerical Integration of User-Defined Function (Simpson's One-Third Rule)* which was utilized for calculating both $G(e_0)$ and $\Gamma(e_0^2)$.

to Lyttleton's, the same additional values of γ are suggested: 0.75, 0.70 and 0.65. These are chosen arbitrarily and conform with the statement above. They depend on the albedo, which is considered to be 0.1, and on the concentration of the reflected light back toward the sun. A closer look shows that the value of γ is weakly dependent on κ when the value of κ is greater than zero. If $\psi = 0.1$, γ approaches 0.80 as κ approaches infinity. For the limiting case, $\psi = 1$ and $\kappa = \infty$, which physically means that all the incident radiation is reflected directly back toward the sun. Working backwards, the expected values of κ can be determined for the corresponding values of γ, which were assumed, using (107). Thus, $\kappa = -1.56$ for $\gamma = .75$, $\kappa = -2.27$ for $\gamma = .70$, and $\kappa = -2.50$ for $\gamma = .65$. It is interesting to note that the light is being reflected more to the sides than back toward the sun for these cases. If the albedo is increased to > 0.4 there will be a concentration of light back in the direction of the sun, *i.e.* κ will be greater than zero. Table IV also gives the values of the integral in (106) for these various values of γ and is represented as $\Gamma(e_0^2)$. Since the function becomes infinite when $y = 0$, a very small finite value must be used as the lower limit rather than zero. In reality, this is quite acceptable since the particle most likely evaporates as it approaches the inner zone of the solar system.

Since the case for which $\gamma = .65$ corresponds to the greatest change in times of infall, this is the value which will be used in these applications. Using the corresponding values of ψ and κ, (106) can be rewritten as

$$t - t_0 = 6.48\times10^6 \; a\rho C^2 e_0^{.3} \int_y^{y_0} y^{-.35}(1-y)^{-3/2}dy \; . \qquad (108)$$

TABLE V

INTEGRAL VALUES FOR EQUATIONS (98) AND (106)

e_o	e_o^2	$G(e_o)$	$\Gamma(e_o^2)$ $\gamma=.80$	$\Gamma(e_o^2)$ $\gamma=.75$	$\Gamma(e_o^2)$ $\gamma=.70$	$\Gamma(e_o^2)$ $\gamma=.65$
.001	.000001	.0000	.0000	.0000	.0001	.0002
.01	.0001	.0004	.0008	.0013	.0023	.0039
.05	.0025	.0052	.0104	.0149	.0216	.0314
.10	.01	.0158	.0316	.0424	.0572	.0776
.15	.0225	.0305	.0612	.0786	.102	.132
.20	.04	.0489	.0980	.122	.154	.195
.25	.0625	.0710	.142	.174	.213	.264
.30	.09	.0969	.194	.233	.281	.340
.35	.1225	.127	.254	.300	.356	.424
.40	.16	.162	.324	.377	.441	.518
.45	.2025	.202	.405	.466	.537	.622
.50	.25	.249	.499	.567	.646	.740
.55	.3025	.305	.609	.685	.771	.874
.60	.36	.370	.705	.824	.918	1.03
.62	.3844	.400	.801	.887	.984	1.10
.64	.4096	.432	.865	.945	1.06	1.17
.66	.4356	.468	.936	1.03	1.13	1.25
.68	.4624	.506	1.01	1.11	1.22	1.34
.70	.49	.548	1.10	1.20	1.31	1.43
.72	.5184	.595	1.19	1.29	1.41	1.54
.74	.5476	.647	1.29	1.40	1.52	1.65
.76	.5776	.705	1.41	1.52	1.64	1.78
.78	.6084	.771	1.54	1.65	1.78	1.92
.80	.64	.846	1.69	1.81	1.94	2.08
.81	.6561	.889	1.78	1.90	2.03	2.17
.82	.6724	.934	1.87	1.99	2.12	2.27
.83	.6889	.983	1.97	2.09	2.22	2.37
.84	.7056	1.04	2.08	2.20	2.33	2.49
.85	.7225	1.10	2.19	2.32	2.46	2.61
.86	.7396	1.16	2.32	2.45	2.59	2.75
.87	.7569	1.23	2.47	2.60	2.74	2.90
.88	.7744	1.32	2.63	2.76	2.91	3.07

TABLE V (cont.)

e_0	e_0^2	$G(e_0)$	$\Gamma(e_0^2)$ $\gamma=.80$	$\Gamma(e_0^2)$ $\gamma=.75$	$\Gamma(e_0^2)$ $\gamma=.70$	$\Gamma(e_0^2)$ $\gamma=.65$
.89	.7921	1.41	2.82	2.95	3.10	3.26
.90	.81	1.51	3.03	3.16	3.31	3.48
.91	.8281	1.64	3.27	3.41	3.56	3.73
.92	.8464	1.78	3.56	3.70	3.86	4.03
.93	.8649	1.96	3.91	4.06	4.21	4.39
.94	.8836	2.17	4.34	4.49	4.65	4.83
.95	.9025	2.45	4.90	5.05	5.21	5.39
.96	.9216	2.82	5.65	5.80	5.97	6.15
.97	.9409	3.37	6.75	6.91	7.08	7.27
.98	.9604	4.30	8.59	8.75	8.93	9.12
.99	.9801	6.37	12.74	12.91	13.09	13.29
.999	.9980	23.091	43.32	43.50	43.69	43.90
.9999	.9998	251.45	140.04	140.22	140.41	140.62

Of course, the same sort of expression can be obtained for the other values of γ. The more restricted form of (108) can now be easily evaluated.

The following example will demonstrate to what extent the inclusion of reflection in the problem affects the time of infall. Take a particle with a mass of 10^{-3} g, which is larger than the particulate mass of two-thirds of the interplanetary dust cloud. Its density will likely be near 1 g/cm^3, which implies that its radius is about 6.2×10^{-2} cm. Let its original semimajor axis be 3 AU. If this particle originated from asteroid collisions, its eccentricity might be fairly low, say 0.5. With the above information, the time of infall can be calculated for both cases. Remembering that $C^2 = a_0^2(1 - e_0^2)^2 e_0^{-8/5}$, the values of the different components of (98) are

$$C^2 = 15.3, \quad ap = 6.2\times10^{-2} \quad \text{and} \quad G(e_0) = .249 .$$

Thus,

$$t - t_0 = 2.6\times10^6 \; years .$$

For the case of partial reflection

$$e_0^{.3} = .812 \quad \text{and} \quad \Gamma(e_0^2) = .740 .$$

Thus,

$$t - t_0 = 3.7\times10^6 \; years .$$

The percent difference for these two times is 35%. This is not a large value in view of the uncertainties involved in estimating the true densities, sizes, and other physical properties of actual particles. In this respect, the times of infall obtained for the restrictive case of total absorption are a good estimation of reality, and until a more precise knowledge concerning specific dust grains is available, this simpler approach will certainly suffice.

Of course, much of the dust in the solar system is left behind by comets in highly eccentric orbits. Assuming all the same values except a new original eccentricity of 0.9, the new lifetimes would be 4.0×10^5 and 5.2×10^5 years for the cases of total absorption and partial reflection respectively. This is a decrease in one order of magnitude which means that comet dust is not only highly susceptible to being blown out of the solar system by radiation pressure as pointed out earlier, but what remains undergoes a much more rapid decay of its orbit due to the Poynting-Robertson effect.

The opposite extreme is mostly interesting in the sense that it places an upper bound on the expected lifetime of a given particle. This is the case of a particle starting in an initially circular orbit.

Assuming $e_0 = 0$, equation (92) becomes

$$a\,da = -\,2\alpha\,dt \ ,$$

and by integrating from 0 to a_0

$$t = \frac{a_0^2}{4\alpha} \ ,$$

or

$$t = 6.98\times10^6\, a_0^2 a\rho \ . \qquad (109)$$

For the case of partial reflection with $\gamma = .65$, the same approach is used with (91). The result is

$$t = \frac{a_0^2}{4\beta} \ ,$$

or

$$t = 9.97\times10^6\, a_0^2 a\rho \qquad (110)$$

since $\beta = .7\alpha$ for this case.

If the same particle is considered again, its time to fall into the sun from a circular orbit will be 3.9×10^6 years for the case of total absorption, and 5.6×10^6 years for the case of partial reflection. It is noted that the increase in the lifetime in changing from .5 to 0 original eccentricity is not nearly equal to the decrease due to the change from .5 to .9. Obviously the particle begins to spiral in much more rapidly as the eccentricity increases past the .5 value.

In order to facilitate the determination of particle lifetimes, the graphs on the following pages provide a means of directly reading this value for a wide range of particles. Values of $a\rho$ are plotted versus time of infall on logarithmic scales. Each graph represents a particular original semimajor axis, and plots are made for six specific original eccentricities using the case of total absorption. The limiting case of an originally circular orbit is shown, and the other extreme, e_0 = .99, is also shown. For these two eccentricities a broken line has also been plotted to show the case of partial reflection in order to provide a visual representation of the magnitude of its effect. The following letters are used as a key for the graphs:

a → e_0 = .99 d → e_0 = .80

b → e_0 = .96 e → e_0 = .50

c → e_0 = .90 f → e_0 = 0.0 .

The value of the semimajor axis starts at .8 AU since the interplanetary dust cloud seems to reach a minimum in this area. The next graph is for a_0 = 1.0 AU, and from there each graph represents an increase of 0.4 AU.

It should be cautioned that the restrictions noted earlier concerning applicable ranges of sizes should be adhered to in reading the time of extinction from these graphs. Radiation pressure or planetary capture may be more effective in removing dust than the Poynting-Robertson effect for extreme values of ap on the graphs. A simple determination of the lower limit can be made by making the assumptions that there are no diffraction effects, and that all the incident radiation is utilized in exerting a pressure force on the particle. Using these assumptions, when the ratio of αc to GM is > 1, the radiation pressure will be greater. This results in a lower limit for the product ap of 5.9×10^{-5}. The upper limit can be determined from the information in Part II.

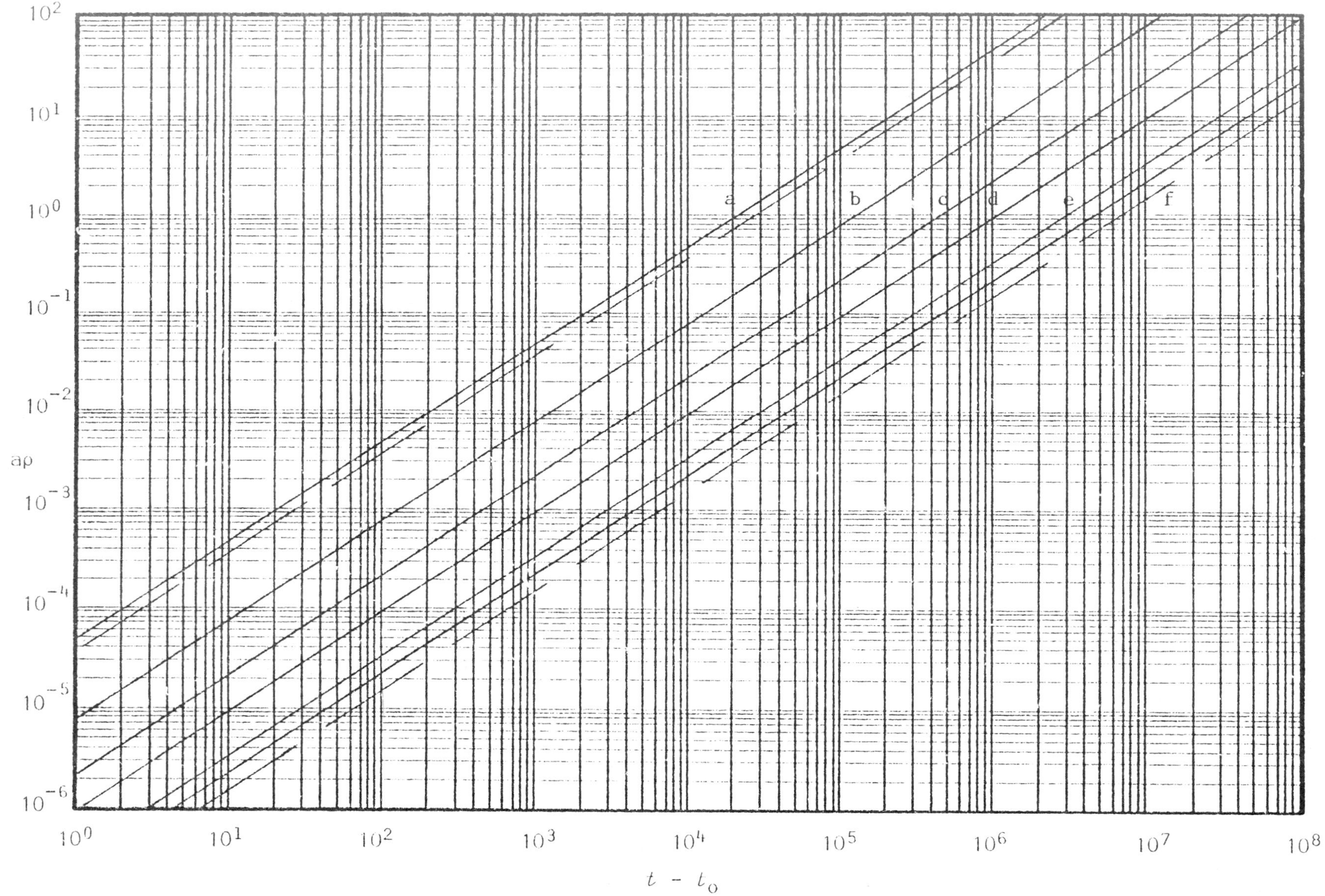

Figure (8). Time of infall for a_o = .80 AU.

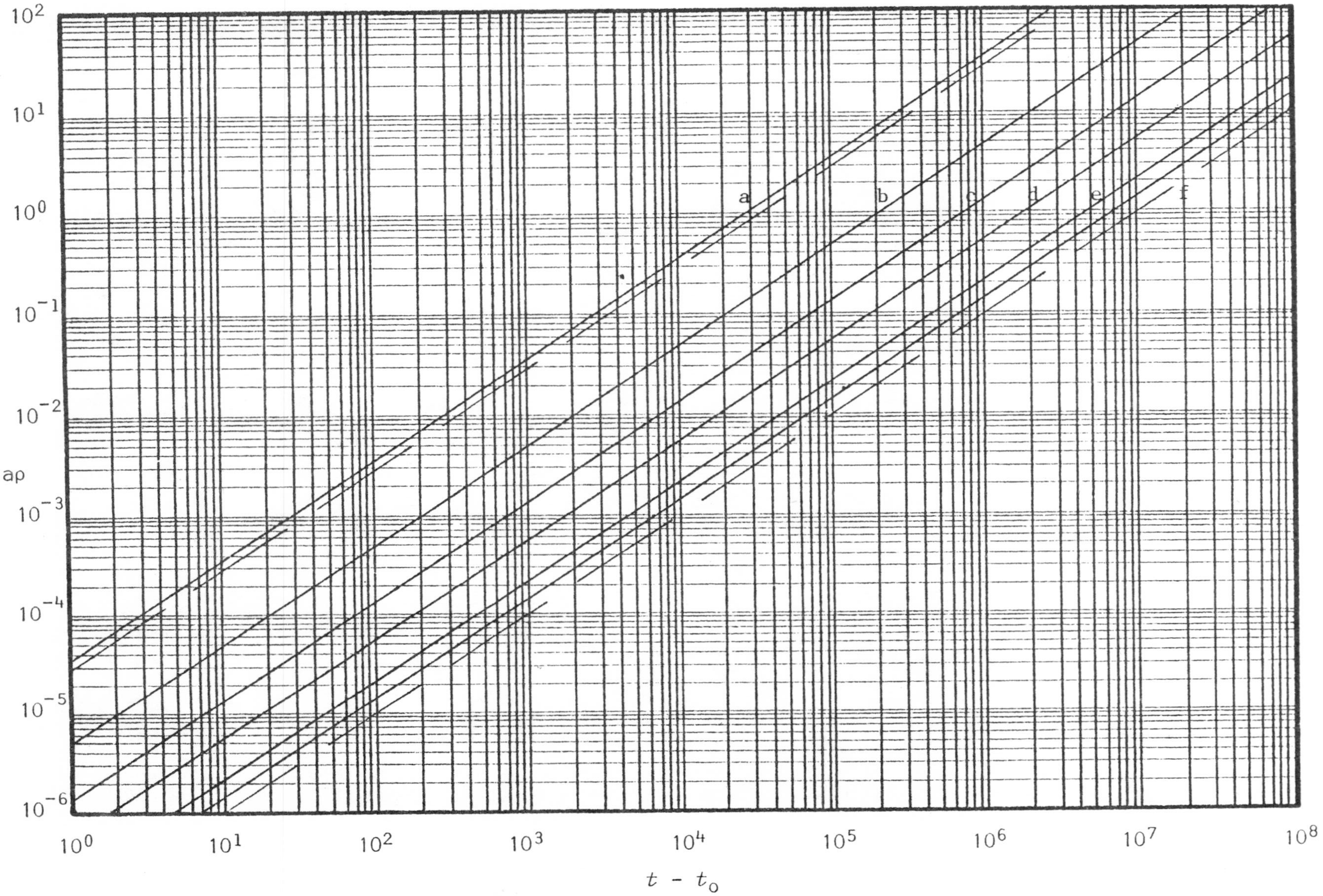

Figure (9). Time of infall for a_o = 1.0 AU.

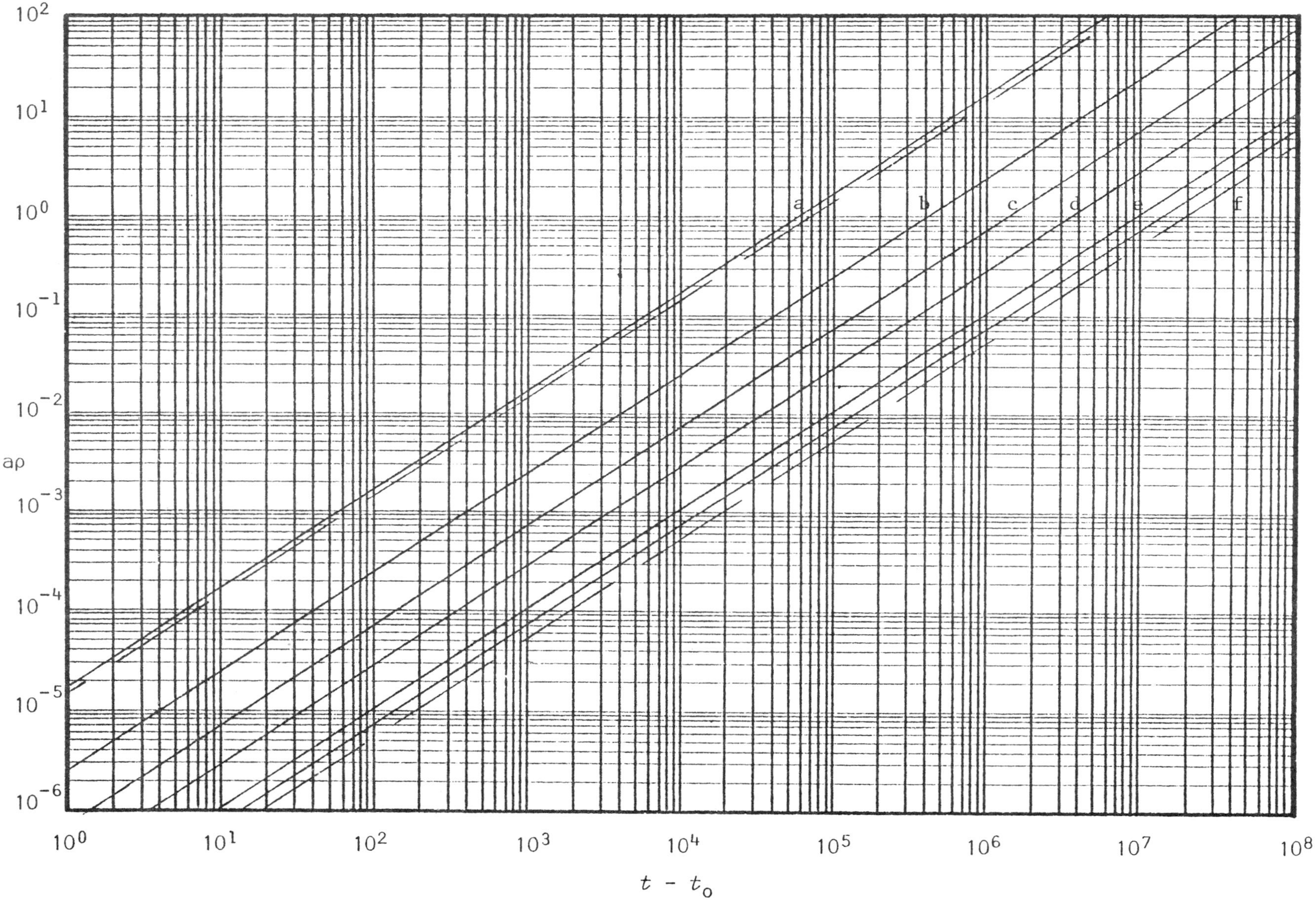

Figure (10). Time of infall for a_o = 1.4 AU.

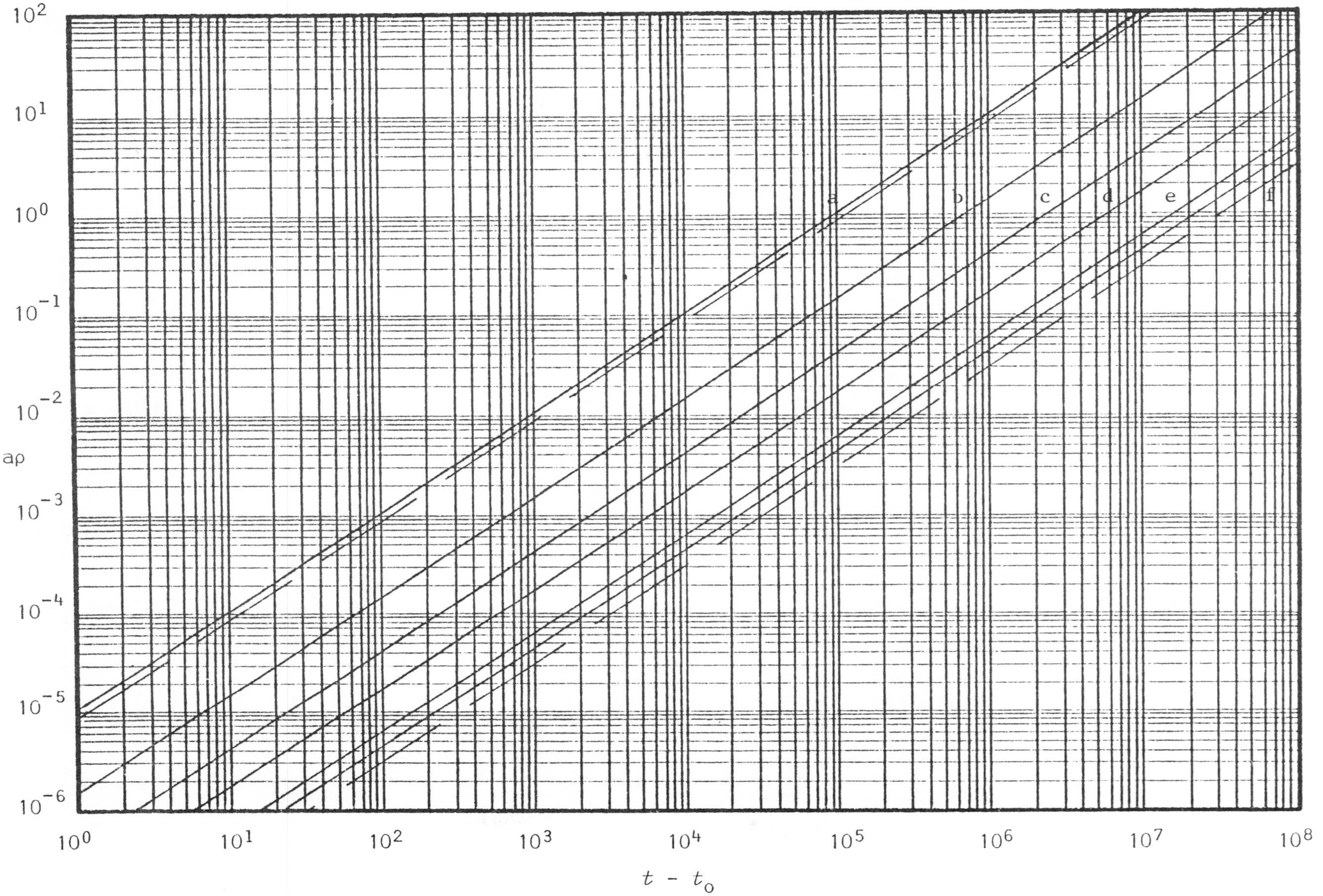

Figure (11). Time of infall for a_o = 1.8 AU.

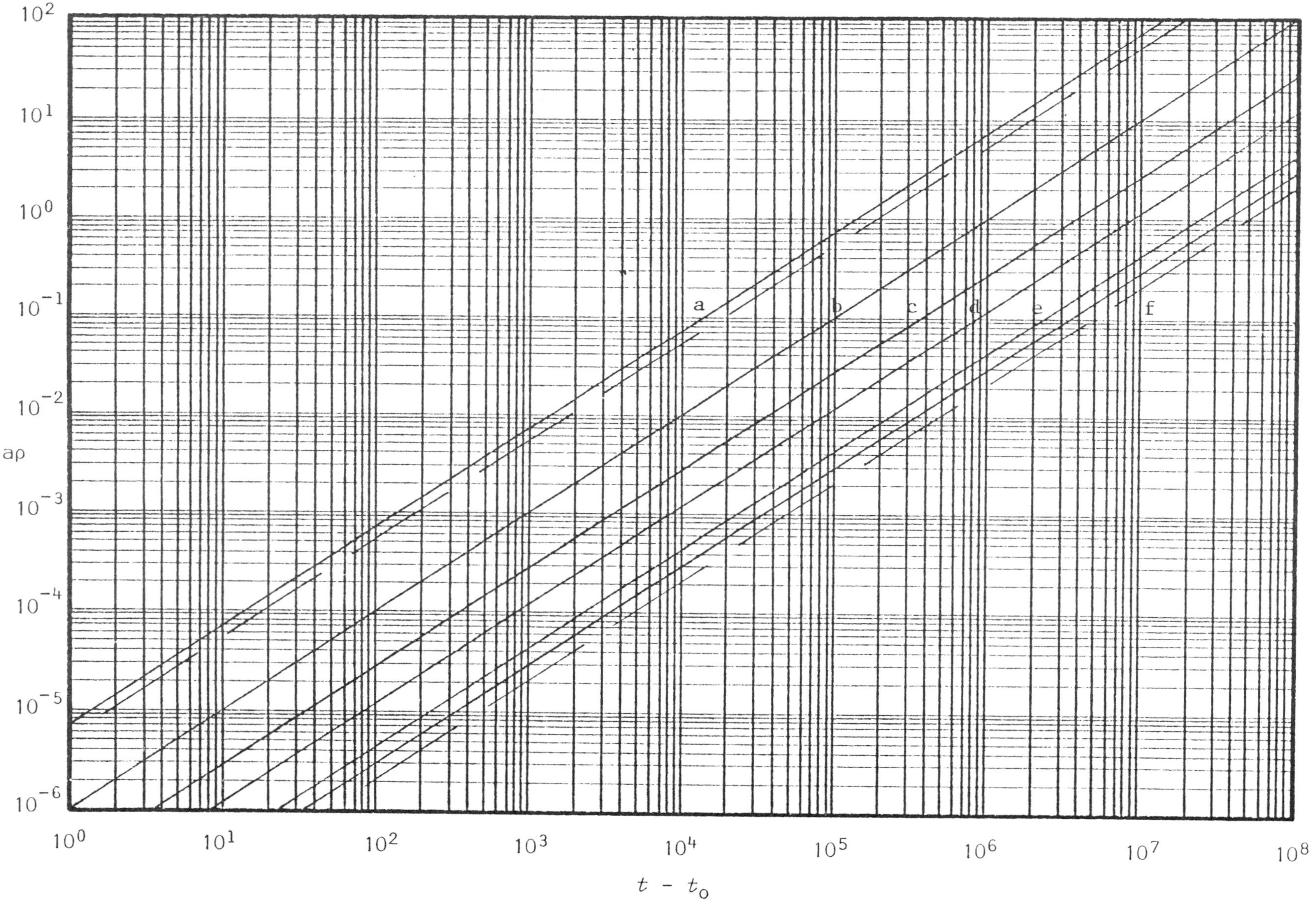

Figure (12). Time of infall for a_o = 2.2 AU.

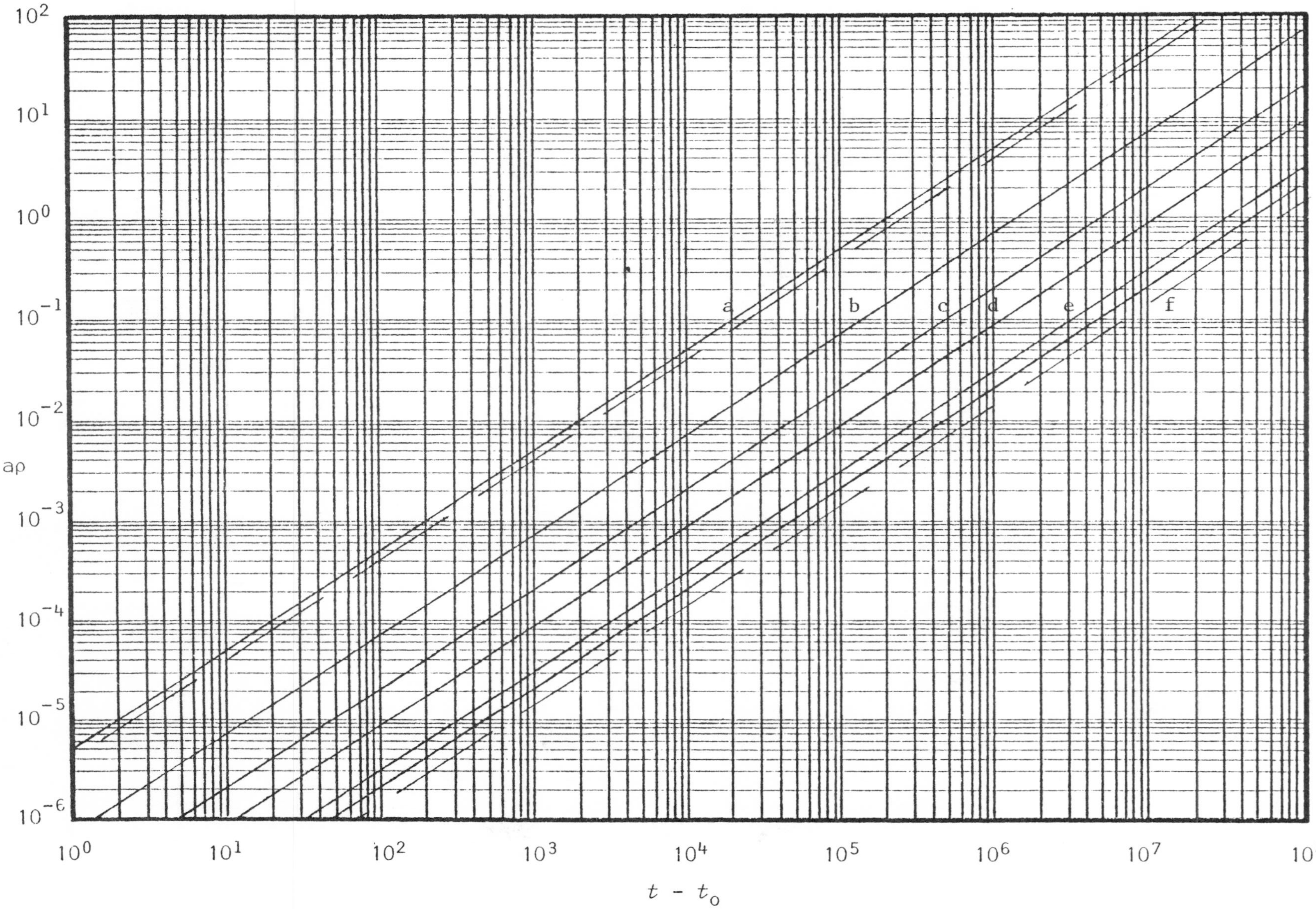

Figure (13). Time of infall for a_o = 2.6 AU.

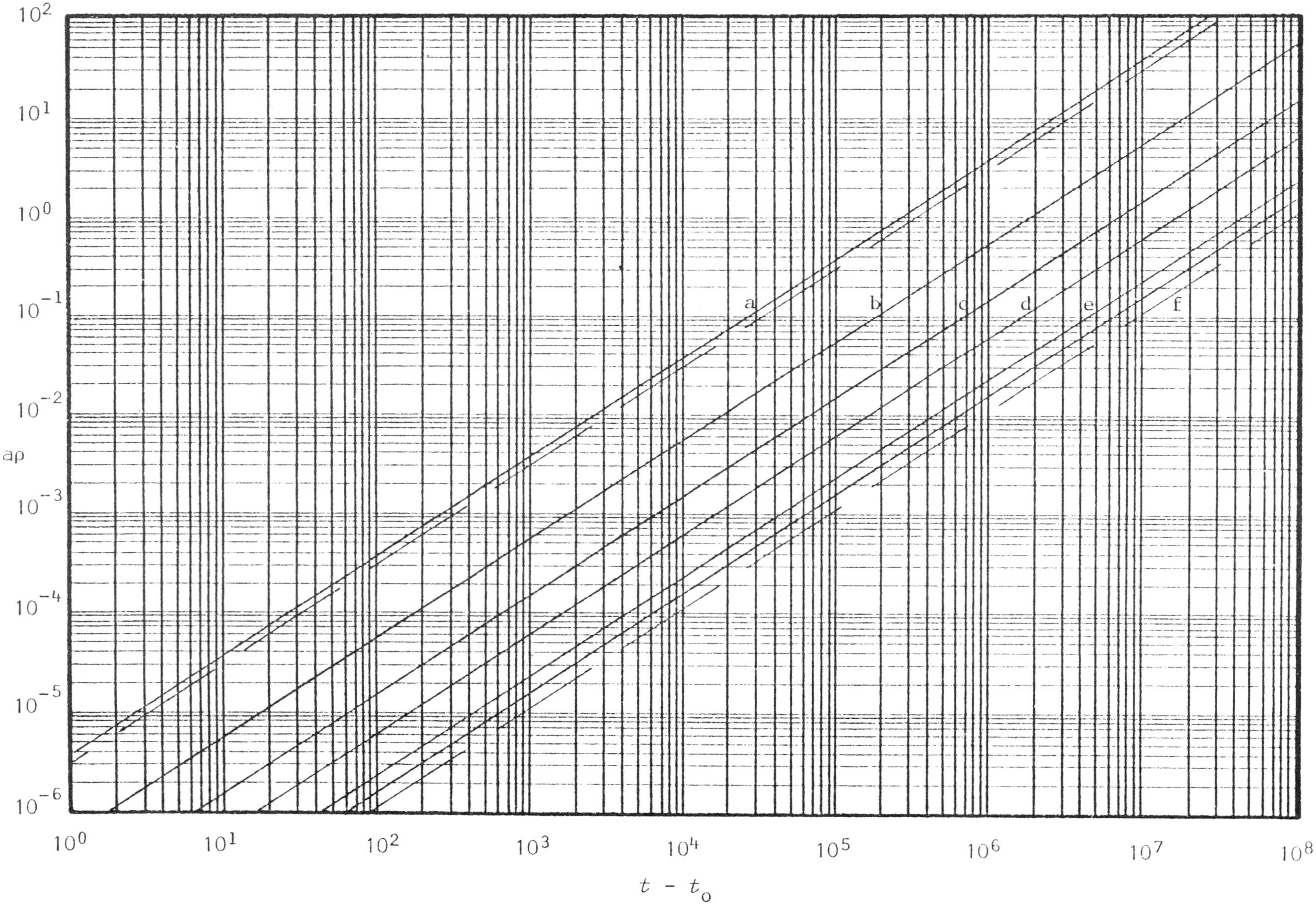

Figure (14). Time of infall for a_o = 3.0 AU.

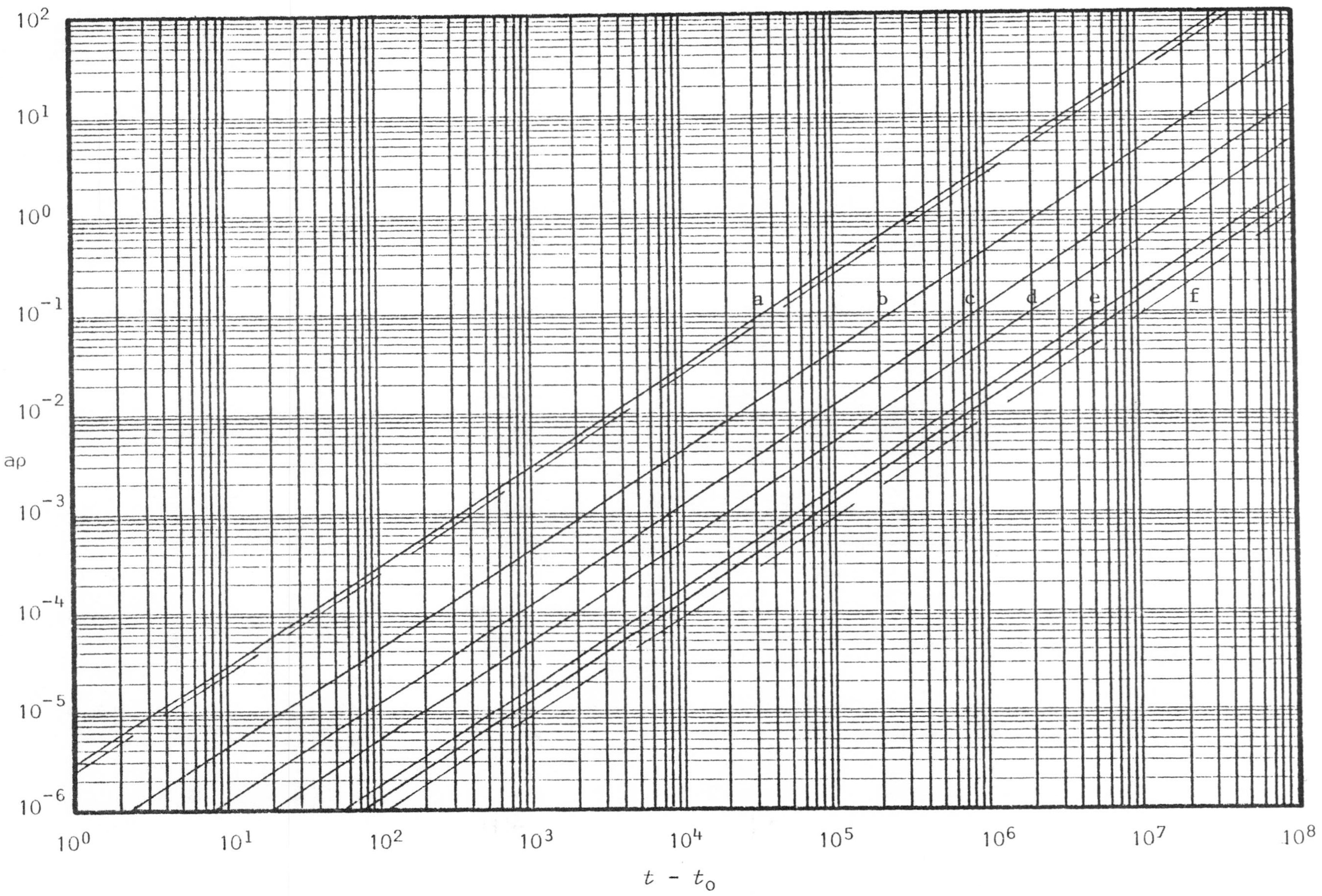

Figure (15). Time of infall for a_o = 3.4 AU.

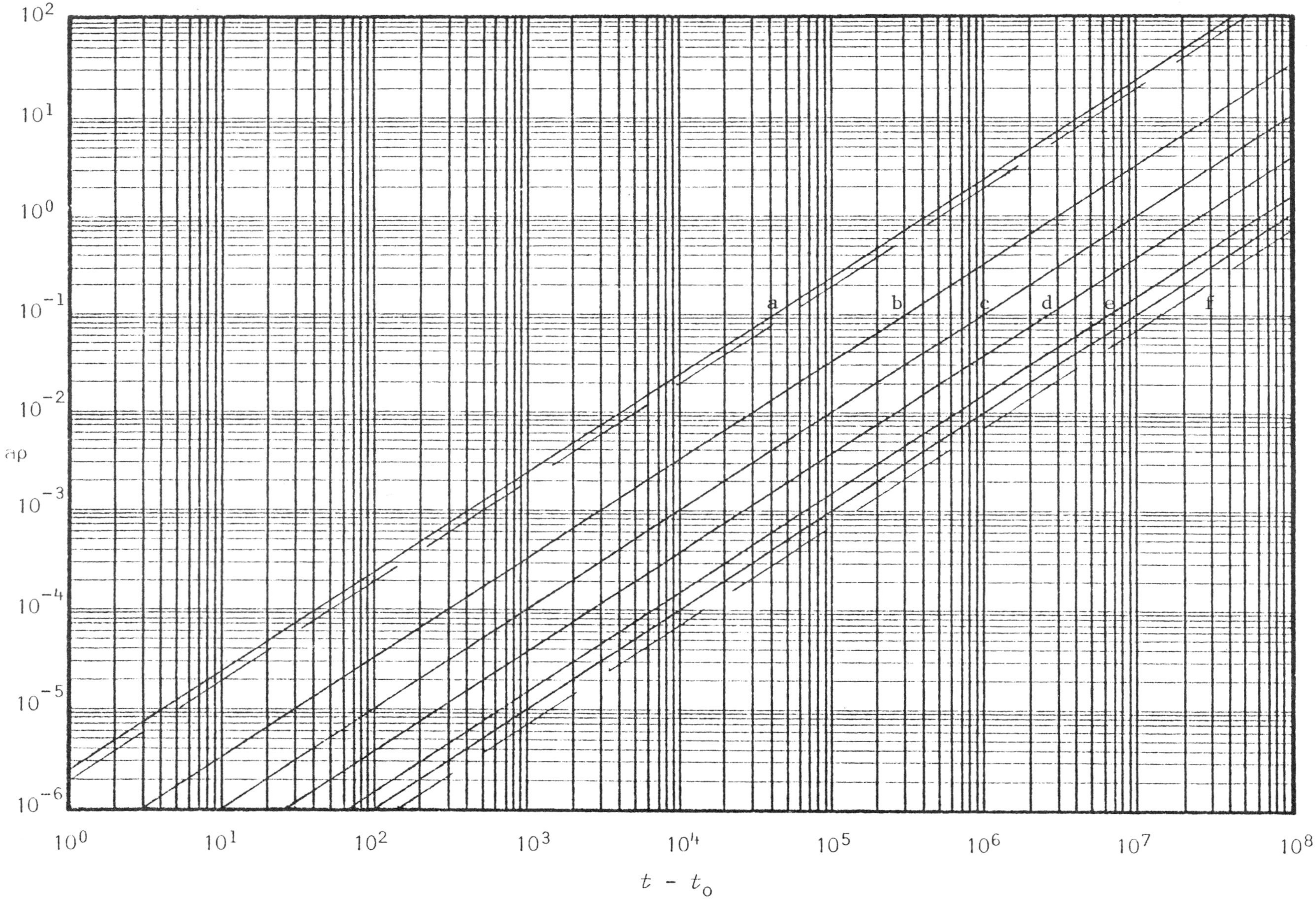

Figure (16). Time of infall for a_o = 3.8 AU.

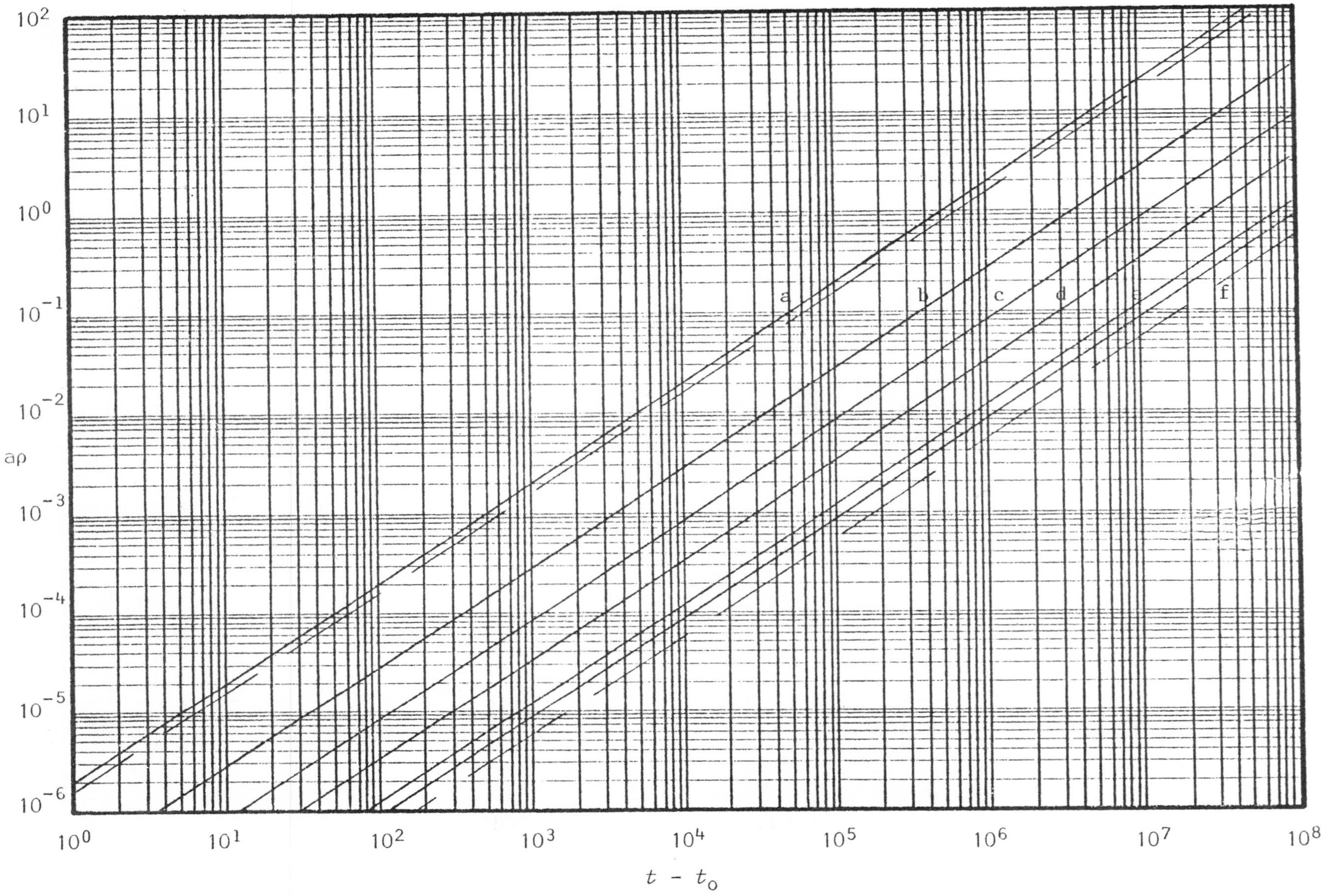

Figure (17). Time of infall for a_o = 4.2 AU.

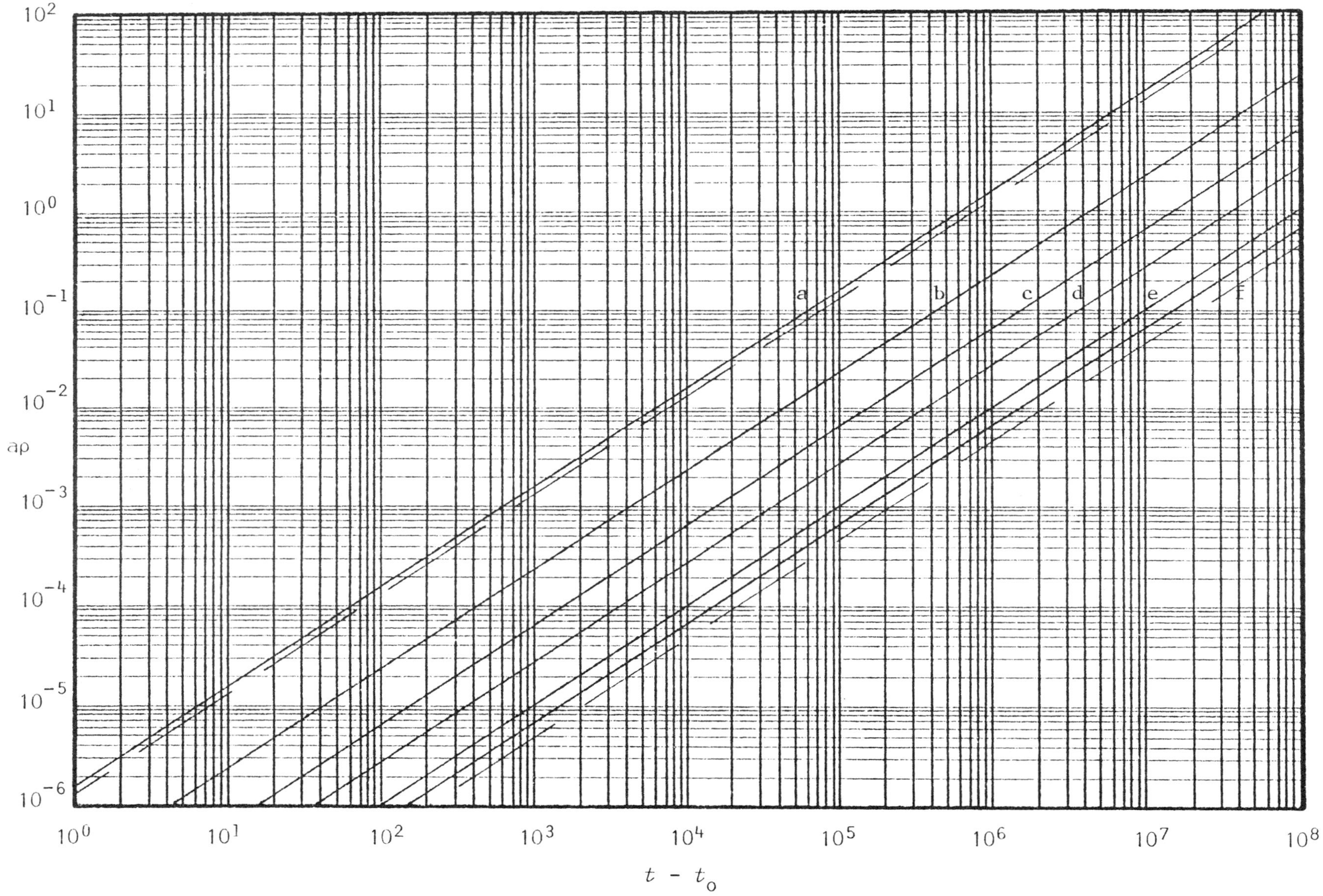

Figure (18). Time of infall for a_o = 4.6 AU.

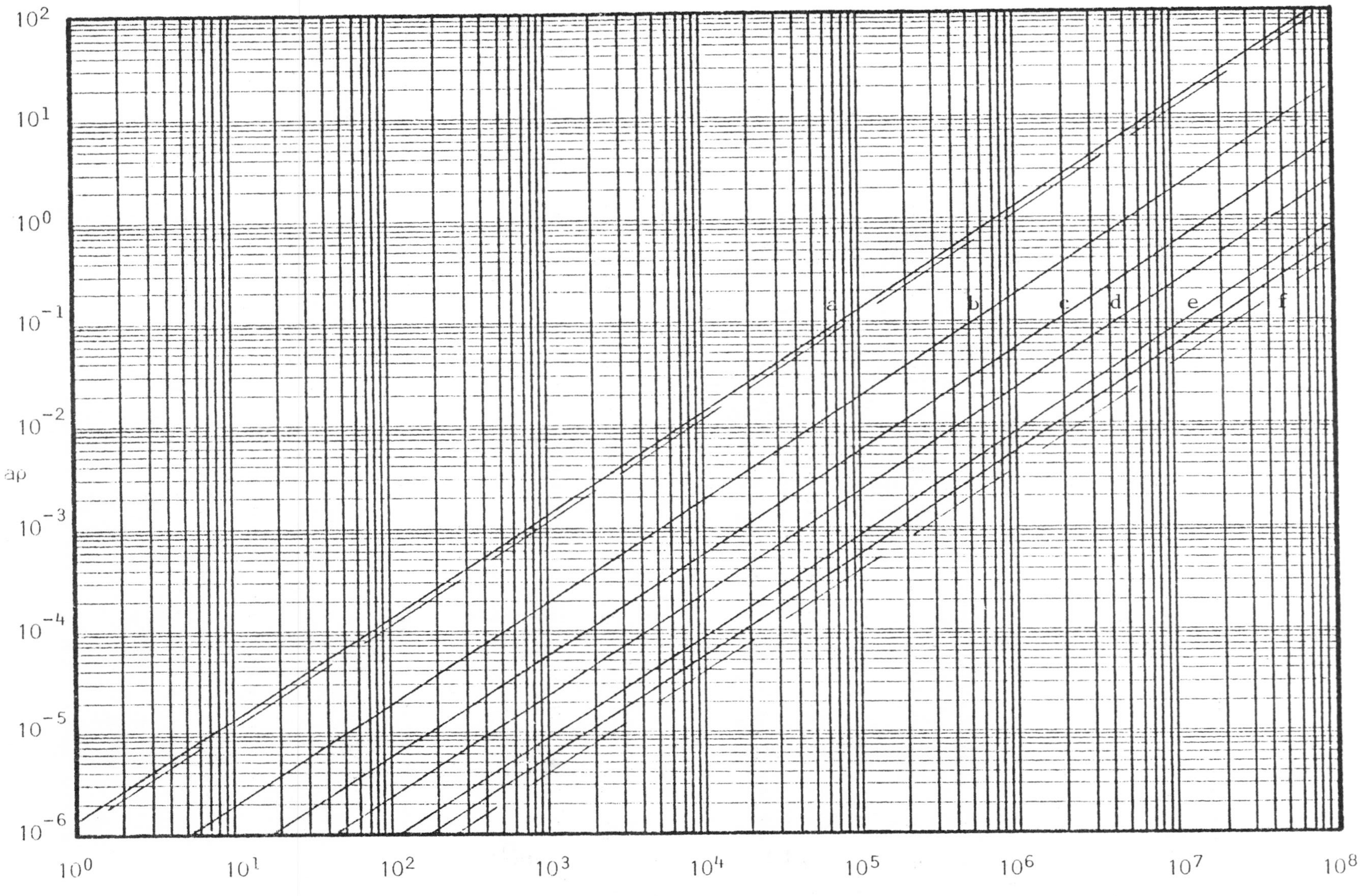

Figure (19). Time of infall for a_o = 5.0 AU.

PART V

CONCLUSION

The Poynting-Robertson effect of solar radiation on small particles in the solar system has been studied in detail from its fundamental assumptions, through the lengthy derivations, to the applications concerning particle lifetimes. The results of this investigation have confirmed the equations reported in various articles since 1937, and have provided evidence that the larger portion of the presently existing solar system dust cannot be expected to survive ten million years. Acting alone, this is a fairly rapid dissipative force. In conjunction with radiation pressure, evaporation, planetary capture, and rotational bursting, the Poynting-Robertson effect is causing the interplanetary dust cloud to undergo a significant erosion which necessitates a considerable source of new material, or dooms the dust cloud to be a short-lived phenomenon with a fairly recent origin as suggested by Delsemme.

None of the mechanisms for resupplying the dust complex in the solar system seem adequate for maintaining the dust complex against the forces of extinction for any time remotely approaching evolutionary guesses on the age of the solar system. The short-period comets which have been suggested as a major source of resupply for the dust complex certainly do not appear as a source for very long since the evidence points to their early demise around ten thousand years at the most after their origin.[45]

If the major portion of the dust and particulate matter presently in the interplanetary medium originated at the same time as the solar system, as many astronomers hold, then the time for spiralling into the sun, because of the Poynting-Robertson effect, puts an upper limit on the age of the solar system vastly less than the evolutionary estimates of its age. In less than two billion years any matter would be swept into the sun that is less than 182.88 cm (6 ft) in diameter inside earth's orbit; less than 7.62 cm (3 in) in diameter inside Jupiter's orbit; and less than 0.225 cm (0.1 in) inside Neptune's orbit. Considering the fact that dust particles much smaller than those are still around in great abundance in the interplanetary medium there would seem to be a maximum to the age of solar system on the order of just several thousand years.

---

[45] H. S. Slusher, "Some Astronomical Evidences for a Youthful Universe," Creation Research Society Quarterly, 8, 1 (June, 1971), p. 57.

APPENDICES

APPENDIX I

PHYSICAL CONSTANTS:

b	1 astronomical unit (AU)	1.496×10^{13} *cm*
c	speed of light	2.998×10^{10} $cm\ s^{-1}$
G	gravitational constant	6.67×10^{-8} $dyne\ cm^{2}\ g^{-2}$
M	mass of sun	1.991×10^{33} g
§	solar constant	1.36×10^{6} $erg\ s^{-1}\ cm^{-2}$

SELECTED SYMBOLS:

a	radius of particle
a	semimajor axis of orbit
e	eccentricity of orbit
f	radiation pressure force acting on particle at rest in S
$\mathbf{k}$	particle reflection vector
l^{μ}	radiation vector in S
m	rest mass of particle
$\mathbf{n}$	unit vector in radial direction
S	inertial system of sun
u^{μ}	velocity of particle in S
α	$2.54\times10^{11}/a\rho$ $cm^{2}\ s^{-1}$
λ^{μ}	radiation vector in Σ_E
μ	$GM - \alpha c$
ρ	density of particle
Σ_E	rest system of particle at instant E
υ^{μ}	particle velocity in Σ_E
ψ	surface albedo
§	unit vector in angular direction
$\mathbf{v}$	velocity of particle

APPENDIX II

MATHEMATICAL NOTATION

The purpose of this appendix is to give a very brief explanation of some of the notation used in this thesis, equation (1) being a prime example. This style of notation was used in this thesis in order to clarify exactly what had been done in the original work by Robertson. This short summary is not intended to be a rigorous treatment of this subject in any way. For a complete presentation of the theory of relativity using this style and approach, the book by Eddington is recommended. This same book may have influenced Robertson in his choice of notation in 1937.

The interval between any two events which takes place in 4-space can be shown to be an invarient. That is, the interval, ds, between them will be the same no matter what reference frame was being used to take measurements of time and space. If the coordinates of the two events are expressed as (x_1, x_2, x_3, x_4) and $(x_1 + dx_1, x_2 + dx_2, x_3 + dx_3, x_4 + dx_4)$, the square of the magnitude of this interval can be expressed as

$$\begin{aligned} ds^2 = {} & g_{11}dx_1^2 + g_{22}dx_2^2 + g_{33}dx_3^2 + g_{44}dx_4^2 \\ & + 2g_{12}dx_1dx_2 + 2g_{13}dx_1dx_3 + 2g_{14}dx_1dx_4 \\ & + 2g_{23}dx_2dx_3 + 2g_{24}dx_2dx_4 + 2g_{34}dx_3dx_4 \ . \qquad (i) \end{aligned}$$

The coefficients will be some function of the coordinates x_1, x_2, x_3 and x_4. Rather than writing this all out every time, (i) can be rewritten as

$$ds^2 = g_{\mu\nu}dx_\mu dx_\nu \ . \qquad (ii)$$

This can be interpretted as indicating that a double summation should be performed over the range of values of μ and ν. This would be expressed mathematically as

$$ds^2 = \sum_\mu \sum_\nu g_{\mu\nu} dx_\mu dx_\nu \ . \qquad (iii)$$

Whenever a literal suffix appears twice in an expression, this operation is simply assumed, making the notation less cumbersome. To give a visual representation, $g_{\mu\nu}$ is a tensor of the form

$$\begin{pmatrix} g_{11} & g_{12} & g_{13} & g_{14} \\ g_{21} & g_{22} & g_{23} & g_{24} \\ g_{31} & g_{32} & g_{33} & g_{34} \\ g_{41} & g_{42} & g_{43} & g_{44} \end{pmatrix}$$

where $g_{\mu\nu} = g_{\nu\mu}$. The product of this tensor with the two vectors dx_μ and dx_ν is expression (*i*).

The symbols $g_{\mu\nu}$ and $g^{\mu\nu}$ also represent a lowering and raising operator. A result of its operation can be to change a covariant vector to a contravariant vector, or vice versa. For example,

$$g_{\mu\nu} dx^\mu = dx_\nu \ . \qquad (iv)$$

In the same manner

$$g^{\mu\nu} dx_\nu = dx^\mu \ . \qquad (v)$$

Applying this, as was done in equation (1),

$$g_{\mu\nu} dx^\mu dx^\nu = dx_\nu dx^\nu \ . \qquad (vi)$$

The expression of the right side of (*vi*), when summed over ν, can be written as

$$dx_1^2 + dx_2^2 + dx_3^2 + dx_4^2 \ .$$

This is the inner product, or dot product, of the two vectors dx^μ and dx^ν. The expression $dx_\mu dx^\nu$ would indicate the outer product of these two vectors. Equation (1) stated that

$$g_{\mu\nu}dx^\mu dx^\nu = dt^2 - (dx^2 + dy^2 + dz^2)/c^2 \ .$$

Applying the lowering operator and performing the implied summation should make the relationship between the right and left sides of this expression readily apparent to the most casual observer. The exact form of this equation is arbitrary. It could also have been written as

$$dx^2 + dy^2 + dz^2 - dt^2c^2 \ .$$

BIBLIOGRAPHY

Abell, George. *Exploration of the Universe.* Second Edition. New York: Holt, Rinehart and Winston, 1969.

Alvarez, J. M. "The Cosmic Dust Environment at Earth, Jupiter and Inetrplanetary Space: Results from Langley Experiments on MTS, Pioneer 10 and 11," in *Interplanetary Dust and Zodiacal Light*, eds. H. Elsässer and H. Fechtig. New York: Springer-Verlag, 1975.

Brandt, John C. and Paul W. Hodge. *Solar System Astrophysics.* New York: McGraw-Hill Book Company, 1964.

Delsemme, A. H. "Can Comets Be the Only Source of Interplanetary Dust?" in *Interplanetary Dust and Zodiacal Light*, eds. H. Elsässer and H. Fechtig. New York: Springer-Verlag. 1975.

Delsemme. A. H. "The Production Rate of Dust by Comets," in *Interplanetary Dust and Zodiacal Light*, eds. H. Elsässer and H. Fechtig. New York: Springer-Verlag, 1975.

Dohnanyi, J. S. "Sources of Interplanetary Dust: Asteroids," in *Interplanetary Dust and Zodiacal Light*, eds. H. Elsässer and H. Fechtig. New York: Springer-Verlag, 1975.

Eddington, A. S. *The Mathematical Theory of Relativity.* Second Edition. Cambridge: Cambridge at the University Press, 1957.

Elsässer, Hans and H. Fechtig, eds. *IAU Colloquium No. 31 on Interplanetary Dust and Zodiacal Light.* New York: Springer-Verlag, 1975.

Fechtig, H. "In-Situ Records of Interplanetary Dust Particles--Methods and Results," in *Interplanetary Dust and Zodiacal Light*, eds. H. Elsässer and H. Fechtig. New York: Springer-Verlag. 1975.

Field, George B. and A. G. W. Cameron, eds. *The Dusty Universe*, Proceedings of a Symposium honoring Fred Laurense Whipple, Oct. 17-19, 1973. New York: Neale Watson Academic Publications, 1975.

Guess, A. E. "Poynting-Robertson Effect for a Spherical Source of Radiation," *Astrophysical Journal, 135* (1962), 855-866.

Kresák, L. and Peter M. Millman, eds. *Physics and Dynamics of Meteors*, IAU Symposium No. 33 at Tratranská Lomnica, Czechoslovakia, 4-9 September, 1967. New York: Springer-Verlag, 1968.

Larmor, J. "Postscript," in *Collected Scientific Papers*, ed. Gilbert A. Shakespear. Cambridge: Cambridge University Press, 1920.

Lamor, J. Proceedings of the 5th International Congress on Mathematics, 1 (1913), 197ff.

Lyttleton, R.A. "Effects of Solar Radiation on the Orbits of Small Particles," Astrophysics and Space Science, 44 (1976), 199-140.

Millman, Peter M. "Dust in the Solar System," in The Dusty Universe, eds. G.B. Field and A.G.W. Cameron. New York: Neale Watson Academic Publications, 1975.

Millman, Peter M. ed. Meteorite Research, Proceedings of a Symposium on Meteorite Research held in Vienna, Austria, 7-13 August, 1968. New York: Springer-Verlag, 1969.

Paddack, Stephen J. and John W. Rhee. "Rotational Bursting of Interplanetary Dust Particles," in Interplanetary Dust and Zodiacal Light, eds. H. Elsasser and H. Fechtig. New York: Springer-Verlag, 1975.

Page, L. Physical Review, 11 (1918), 376ff.

Page, L. Physical Review, 12 (1918), 371ff.

Poynting, J. H. "Radiation in the Solar System: Its Effects on Temperature and Its Pressure on Small Bodies," Philosophical Transactions of the Royal Society of London, A, 202 (1903), 525-552.

Robertson, H. P. "Dynamical Effects of Radiation in the Solar System." Monthly Notices of the Royal Astronomical Society, 97 (April, 1937), 423-438.

Singer, S. Fred. "Interplanetary Dust," in Meteorite Research, ed. P. M. Millman. New York: Springer-Verlag, 1968.

Singer, S. Fred and Lothar W. Bandermann. "Nature and Origin of Zodiacal Dust," in The Zodiacal Light and the Interplanetary Medium, ed. J. L. Weinberg. Washington, D.C.: NASA SP-150, 1967.

Slusher, H. S. "Some Astronomical Evidences for a Youthful Solar System," Creation Research Society Quarterly, 8, 1 (June, 1971), 55-57.

Vanysek, Vladimir. "Dust in Comets and Interplanetary Matter," in Interplanetary Dust and Zodiacal Light, eds. H. Elsasser and H. Fechtig. New York: Springer-Verlag, 1975.

Weinberg, J. L., ed. The Zodiacal Light and Interplanetary Medium, Proceedings of a Symposium, Jan. 30 - Feb. 2, 1967, Honolulu, Hawaii. Washington, D.C.: NASA SP-150, 1967.

Whipple, Fred L. "On Maintaining the Meteoritic Complex," in The Zodiacal Light and the Interplanetary Medium, ed. J. L. Weinberg. Washington, D.C.: NASA SP-150, 1967.

Wyatt, Stanley P. and Fred L. Whipple. "The Poynting-Robertson Effect on Meteor Orbits," Astrophysical Journal, 111 (1950), 134-141.